Working with
CHEMISTRY

Working with
CHEMISTRY

A LABORATORY INQUIRY PROGRAM

SECOND EDITION

Donald J. Wink

University of Illinois at Chicago

Sharon Fetzer Gislason

University of Illinois at Chicago

Julie Ellefson Kuehn

William Rainey Harper College

W. H. Freeman and Company • New York

Publisher: Susan Finnemore Brennan
Acquisitions Editor: Clancy Marshall
Marketing Manager: Mark Santee
Editorial Assistant: Jenness Crawford
Project Editor: Vivien Weiss
Photo Editor: Vikii Wong
Cover and Text Designer: Diana Blume
Illustration Coordinator: Bill Page
Composition: Techbooks
Production Coordinator: Susan Wein
Manufacturing: RR Donnelley & Sons Company

Library of Congress Cataloging-in-Publication Data
Wink, Donald J.
    Working with chemistry : a laboratory inquiry program / Donald J. Wink,
Sharon Fetzer Gislason, and Julie Ellefson Kuehn.–2nd ed.
        p. cm.
Includes bibliographical references and index.
ISBN 0-7167-9607-4 (pbk.)
1. Chemistry–Laboratory manuals. I. Fetzer-Gislason, Sharon. II. Kuehn,
Julie Ellefson. III. Title.
QD45.W75 2004
542–dc22

                                    2004040352

W. H. Freeman and Company
41 Madison Avenue, New York, NY 10010
Houndmills, Basingstoke   RG21 6XS, England
www.whfreeman.com

Contents

Experiment Groups

Custom Publishing

In the prior edition, we offered "laboratory separates" as an option to purchasing the entire text. For this new edition, we our proud to provide instructors the opportunity to custom publish their own laboratory manual, using any of the experiments presented within this complete manual.

The W. H. Freeman Custom Publishing Program lets you:

1. Create your own printed lab manual, containing only the experiments you intend to cover.
2. Include experiments from this manual or any of the lab manuals published by W. H. Freeman and Company.
3. Continue to use experiments from prior editions, which may have been deleted from this manual.
4. Insert your own experiments and material.
5. Allow blank pages for note taking.

Choose any experiment from:

- *Working with Chemistry*, by Donald J. Wink, Sharon Fetzer Gislason, and Julie Ellefson Kuehn
- *General Chemistry in the Laboratory*, by James M. Postma, Julian L. Roberts, Jr., and J. Leland Hollenberg
- *General, Organic, and Biochemistry Lab Manual*, by Sara Selfe
- *Modern Projects and Experiments in Organic Chemistry*, by Jerry R. Mohrig, Christina Noring Hammond, Paul F. Schatz, and Terence C. Morrill
- *Lab Experiments in Introductory Chemistry*, by Philip Reedy, Donald J. Wink, and Sharon Fetzer Gislason

To create and order your custom book, go to *www.whfreeman.com/custompub*. After registering and logging in, you can complete the entire process—from selecting contents to creating the cover to ordering desk copies. Within four to six weeks of placing the order, your customized books will be printed, bound, and shipped to your bookstore.

For more information on custom publishing:
Contact your W. H. Freeman Sales Representative
Visit *www.whfreeman.com/custompub*
Send a message to *custompub@whfreeman.com*

Preface

The Idea of *Working with Chemistry*

We are pleased to introduce you to the second edition of our general chemistry laboratory program, *Working with Chemistry* (WWC). This program started from the idea that we could build a program for all students that relates general chemistry to people who use chemistry on a regular basis, even if "chemist" is not part of their title.

We do this while covering essential principles taught in general chemistry. Here are just three examples of how standard general chemistry questions are recast in our program:

1. "How does reaction stoichiometry matter in this lab?" becomes "What stoichiometry is used when a field ecologist needs to determine whether a soil is rich in nitrogen?"
2. "What are the characteristics of an acid–base buffer?" becomes "How can a buffer be restored to a patient after cardiac arrest?"
3. "What is the heat capacity of a metal?" becomes "Which of these substances is best suited to protect valuable documents from fire?"

The ideas that underlie these labs are drawn from the education research and policy literature of the last ten to twenty years. The way to forge connections between professional and basic science training has already been enunciated in an NSF report*:

- Form instructional teams of specialists across traditional departmental lines.
- Teach basic science and mathematics through the vehicles of real-world technical problems and industrial scenarios.
- Encourage the collaboration of industrial and academic scientists and technicians.

The WWC materials incorporate these connections in several ways. First, we build upon a discussion of the importance of chemistry in a professional environment with a researcher in that area. Second, each experiment group provides instruction in one or more basic ideas or techniques applied to a genuine scenario. Third, collaborative work is built around individual responsibilities in each experiment.

Goals and Structure of *Working with Chemistry*

The *Working with Chemistry* program has a primary goal that is present in any general science course: understanding how chemistry gathers evidence and solves problems. To accomplish this, we have two additional goals that will serve this primary goal. First, students should develop a mastery of techniques and ideas so that they can gather and interpret data independently. Second, the students should be able to articulate, with reference to the written materials and personal experience, how someone

---

*National Science Foundation. *Gaining the Competitive Edge: Critical Issues in Science and Engineering Technician Education* (NSF 94-32). National Science Foundation: Washington, DC, 1994.

in a real work environment might find a particular technique or idea useful in solving a problem.

Note that these goals are *not* vocational. A general chemistry course should not be viewed as a place to learn particular technical skills. But *Working with Chemistry* does aim to give students strong academic skills that will be remembered when it is time to learn actual workplace skills. This is why we feel the WWC program is also important to students in normal academic tracks. For them, an enhanced understanding and enjoyment of chemistry should aid in the development of a clear understanding of the importance of basic science.

Each *Working with Chemistry* experiment group is written to cover two weeks of experimentation. Students read the scenario before the first week's work, and this anchors the instruction for the entire experiment group. The scenario returns as the focus of the important problem they are asked to address in the second experiment.

Skill-Building Labs show students how to use a technique. These labs are largely drawn from existing methods commonly taught in the General Chemistry Laboratory.

Application Labs return to the professional scenario we have identified. They are more open in their inquiry style.

The Relevance of *Working with Chemistry*

The sheer number of majors that require general chemistry (25 at our schools alone) make it one of the most notorious courses on any campus. Students struggle with the inherent difficulty of the cumulative learning required in chemistry and their need for chemistry to prepare themselves for a distant employment goal. Orville Chapman of UCLA has clearly (and bluntly) summed up how this struggle compromises the effectiveness of a traditional fact- and computation-focused curriculum:

> We have no hope of expanding our clientele with our present structure. The structure is not sound. The message from 80% of college students comes in loud and clear. Chemistry without people, economics, and policy is irrelevant. We choose to be irrelevant; they ignore us.*

This call for relevance can be and is addressed in curricula that make students more aware of the importance of chemistry to societal problems, especially those of the environment. Another approach is to bring more contemporary research problems into the course. Both have the advantage of making "newspaper" and "community" chemistry accessible to the student. However, for students more focused on a professional degree there needs to be something more: a clear demonstration that, yes, chemistry does matter to *employment*.

In our case, we felt that the laboratory is the most attractive place to link learning and application. The *Working with Chemistry* labs allow students to experience how the solution of real problems by people in all walks of life requires a thorough understanding of general chemistry principles.

Interdisciplinary Instruction in *Working with Chemistry*

The WWC program includes a strong interdisciplinary basis that, we hope, provides meaningful issues, examples, and experiences from outside of chemistry. Certainly as authors we have been enriched in the course of discussions in which we learned, for

---

*National Science Foundation. *Innovation and Change in the Chemistry Curriculum* (NSF 94-19). National Science Foundation: Washington, DC, 1993.

example, how equilibria are used in analyzing protein–drug interactions or the way that heat capacity of materials is important in protecting against fire. The same enrichment occurs when students construct a calibration curve to assay albumin and compare results about their "patients."

As suggested earlier, we worked hard to structure the interdisciplinary approach of *WWC* so that it did not compromise the core subject, chemistry. That is critical in classes where students often have varying skill levels before the class starts. We carefully wrote the skill-building and application labs to bring all students to a common understanding of how the chemistry of the system works before we ask them to think about using chemistry in another area.

Inquiry, Teaching, and Learning in *Working with Chemistry*

Each experiment group connects to one or more of the *concepts and skills in general chemistry*. This keeps in place the "standard" lab goal of supporting basic fact and concept learning. Also, it provides adequate training during the first two weeks of an experiment group to allow students to appreciate fully the relevance of the applications in the second week. When we address the problem of providing a *working environment*, we do not just want to tell students how chemistry is used by professionals. We want to make them feel the actual environments.

In our implementation of a *guided-inquiry* format, students are responsible for collecting and interpreting data in order to answer some testable question. We do not expect students, for example, to be able to design an acid–base titration experiment on the first day they use a buret. But within a short time span, they will be expected to carry out the procedure to meet certain parameters. In some cases, students determine how to collect the data. They also decide if their data are acceptable and, if not, they have the opportunity to make adjustments and try again. Finally, they are expected to make connections between the scenario, the data they collected, and the chemical principle underlying the experiments.

Professionals are expected to be able to do *collaborative work* from a base of independent competence. Students have ample opportunities to develop individual laboratory skills and to analyze and communicate their own results. However, each experiment group also includes a team component. This may mean collaborating on a procedure, sharing group data, or preparing valid samples for one another.

Acknowledgments

There are only three listed authors of this lab manual, but it represents the collaboration of more than a dozen people who ably and enthusiastically contributed to the program. Several people have played a role in shaping such large sections of the work that they deserve general thanks.

Perhaps the most important coworker for us is Mr. William Haney, Coordinator of the UIC Undergraduate Chemistry Laboratories. Only Bill could ensure us that the kinds of experiments we planned would be practical in real teaching environments. More important, he had a special eye for understanding that what we wrote might not be what the students would understand, and his feedback from classroom observations and conversations with teaching assistants has been a big part of the *WWC* instructor's manual.

Most *WWC* experiment groups have a scenario that was written in conjunction with a professional faculty member or another person thoroughly familiar with the application being simulated. The work was not restricted to just "their" experiment, however. Group discussions and planning by all *WWC* authors and professional faculty offered everyone the opportunity to bounce ideas around freely, thus they all deserve some general thanks for their work. These faculty members are listed in the acknowledgments section of the Introduction.

Drs. Luke Hanley and Audrey Hammerich of the UIC chemistry department also played an internal role as editors of some of the experiment groups. Colleagues who have also taught these same students enabled us to check that our feel for students' prior and developing knowledge was reasonable. Similar outside perspectives were provided by Dr. Allan Smith of Drexel University and Dr. Baird Lloyd of Miami University Middletown. In between the first and second editions, we also worked twice in a workshop, called "Supporting Student Laboratory Learning," funded by a subcontract from Georgia State's Collaborative Workshops in the Chemical Sciences. We want to thank the participants and, especially, Dr. Jerry Smith, the P.I. for the program. Our work in that effort, and in revising some of the labs, was supported by assistance from Dr. Ying Han and Ms. Jennifer Barber-Singh.

The final group of people who contributed in detail to the structure of the WWC labs includes several dedicated colleagues at other schools. First among these are six faculty at Chicago-area community colleges, who helped plan, evaluate, and serve the initial implementation of these labs. They are Drs. Helene Gabelnick and Dennis Lehman (Harold Washington College of the City Colleges of Chicago), Susan Shih and Dan Fuller (College of Du Page), Jerry Maas (Oakton Community College), and Barbara Weil (William Rainey Harper College). In addition, several people have tested these labs over the last year on their own campuses through our initial download option. Their comments on what happened when the WWC labs went to distant sites have helped in particular and in general.

And, of course, the students at our campuses were the ones who first experienced these labs. Their experiences, in part observed and reported upon by Prof. Elizabeth Whitt (Iowa) and Ms. Han Mi Yoon (UIC), were the primary piece of evaluation data for the shaping and reshaping of the WWC labs. A select group also served as special testers, and we proudly share their data in the student results section of the instructor's notes.

Special thanks are due to the many instructors who reviewed this material before publication. Robert Allendoerfer, State University of New York, Buffalo; Susan Bates, Ohio Northern University; Stacy Lowery Bretz, University of Michigan—Dearborn; Dave Bugay, Kilgore College; Dave Cohen, J. Sargeant Reynolds Community College; Alan Cooper, Worcester State University; David Frank, California State University, Fresno; Robert Hammond, University of Maryland, College Park; Rick Hartung, University of Nebraska, Kearney; Sara Iaccobucci, Tufts University; Sanda Lamb, University of California, Santa Barbara; Eric Mechalke, Casper College; Lynne O'Connell, Boston College; Norb Pienta, University of North Carolina, Chapel Hill; Laurence Rosenheim, Indiana State University; Theodore Sakano, Rockland Community College; Marcy Towns, Ball State University; MaryJon Whittemore, Forney High School, Texas; Marie Wolff, Joliet Junior College; Corbin Zea, Creighton University; Gail Zichittella, Cheektowaga Central High School, New York.

We want to thank the many people at W. H. Freeman and Company who helped make this second edition possible, with special thanks to Mark Santee, Todd Elder, Clancy Marshall, and Susan Finnemore Brennan. We also acknowledge the support of the National Science Foundation in getting this project underway and the key connection made by Brian Coppola of the University of Michigan at his Day 2-to-40 symposium in May, 1997.

Finally, we acknowledge the National Science Foundation Division of Undergraduate Education, which gave us the funds to start this project. We proudly include their logo as an indication of our gratitude. The material in this text was created, in part, through support from Grant No. NSF-DUE 9653080.

Any opinions, findings, and conclusions or recommendations in this material are those of the authors and do not necessarily reflect the views of the NSF.

Introduction

I. Program Components

A. Introduction to the *Working with Chemistry* Program

During this semester, you will use one or more experiment groups developed as part of a program called *Working with Chemistry* (WWC). This program is designed to show how general chemistry principles are used by people in many different fields. This includes a large group of people, ranging from field ecologists to chemical engineers, and many health professionals as well.

Working with Chemistry is a program developed by chemists working with faculty from other departments who use chemistry in their everyday work. They may be biologists, pharmacists, or engineers. We refer to them as *chemical professionals*, those who use chemistry in their professional lives. The purpose is not to train you how to be one of these persons, but to show you how professionals use the principles of general chemistry in their work. We hope you will develop a better understanding of how chemistry is used today.

Of course, chemical professionals usually do not work in an environment where the same procedures can be used every time. They usually have to find out something about a patient, for example, and then design a response. This leads us to the second important feature of the *WWC* labs: They are based on your own responsibility to inquire, interpret, and act upon results.

Finally, chemical professionals may be individually responsible for their own work, but they do often work as members of teams. When directed, you will therefore assemble a set of data with a group of students. This pooled data will allow you to carry out your individual work in a faster and more accurate manner.

The design of a *WWC* experiment group is simple. During the experiment group, you will learn techniques that you will then apply in the study of a particular problem. We have drawn the "problems" from the work that is done by chemical professionals, and these people have helped us design each experiment group from the ground up. During an experiment group, you will have more and more responsibility for designing the actual procedures. Thus, the first week is the "Skill-Building Lab" and the second week is the "Application Lab."

B. Forming and Working in Groups

If you have never worked in a group before, it can feel threatening the first time because you may not know what is expected of *you* as an individual. The process is actually very simple, and as you become familiar with group work within a chemistry lab environment, you will realize that it is the same process you often use to approach your everyday tasks. *Each person in the group is important, because each person has a definite part to contribute to the completion of the experiment*. First you must understand what the group is expected to complete by the end of the lab period. This information is contained in the experimental description and procedure, which should be read before coming to lab. In reading through the experiment, you should understand the reasons (or theory) behind the procedure as well as how to perform the necessary measurements. It helps to highlight important information in each experiment before you come to lab.

After your group has listed all the tasks that must be accomplished, these should be distributed among the individual group members. For the experiment to be successfully completed, each group member should clearly understand the assigned task and how it is to be completed. It is always a good idea to *discuss this as a group first* before going ahead with individual tasks. The group discussion should include the degree of accuracy required for all measurements, a review of laboratory techniques, an investigation of new laboratory procedures, and any cautions to be observed within the performance of the experiments. Your group must also organize data collection.

As an example of how the group might work, let's see how these general directives apply to a procedure to study the relative amounts of materials needed in a chemical reaction. If a group of four must report results for two solutions, then two students would study each solution. All members would work with the same volume of the solution, however. That way everyone has the same volume to compare.

Your group may prefer a different division of labor that will accomplish the experimental goals just as well as this one: The essential point is that all work is done and reviewed in the time allotted, with all the members of the group participating. Important things to notice and record are the names of your group members and the letter (or code) of any unknowns assigned to each person. Take time to record this information at the beginning of the lab period.

Members of a well-functioning group should be able to ask questions within their group, provide direction to members in need of such, and both give and accept critical analysis of the quality of their measurements. At the end of the experiment, you may be asked to evaluate the group's performance. This should improve both individual and group understanding and performance.

II. Safety in the Chemical Laboratory

A. Chemical Hazards: Identification and Management

Each person in a laboratory is responsible for maintaining a safe environment. Therefore, you must be aware of any hazards associated with the chemicals used in the experiments. You must always wear appropriate safety equipment, including safety goggles, while working in the lab. It is also your responsibility to follow the safety rules and regulations established by your institution and your laboratory instructor.

Some of the chemicals used in some of the *WWC* experiments have been identified as harmful or potentially harmful to humans. The chemicals may be corrosive, chemicals that destroy living tissue and equipment on contact; flammable or combustible, substances that give off vapors that can readily ignite under usual working conditions; irritants, substances that have an irritant effect on skin, eyes, respiratory tract, and so on; toxic, substances that are hazardous to health when breathed, swallowed, or in contact with skin; carcinogenic, substances that are known or suspected to cause cancer; mutagenic, chemical or physical agents that cause genetic mutations; and teratogenic, substances that cause the production of physical defects in a developing fetus or embryo. To minimize your risk, it is essential that you handle these materials properly, carefully, and while wearing appropriate protective equipment. In fact, a good general rule for all chemicals, even those considered nonhazardous, is to avoid contact with skin, clothing, and eyes. You must also avoid breathing vapors.

Persons who have medical conditions, or may be pregnant, should consult their doctor before participating in any lab course that uses chemicals.

To assist you in understanding the hazards associated with specific chemicals, each *WWC* experiment contains Cautions to alert you to potential hazards. Additional information on each chemical is available on its Material Safety Data Sheet (MSDS). Every MSDS contains specific information including physical and chemical characteristics, health effects, fire and explosion data, reactivity hazard data information, health hazard data, and precautions for spills and cleanup. MSDS sheets should be available from companies that supply reagents. Those in charge of the laboratory should have these on file.

B. Disposal of Chemical Substances

All chemical waste must be disposed of properly, generally in appropriately labeled containers. Your institution has a chemical hygiene plan that includes procedures for waste disposal. Follow all disposal procedures as outlined by your instructor. Be sure to also clean all glassware, equipment, and your work area before leaving the laboratory.

III. Instruments for Chemical Measurement

A. Handling Chemicals

As outlined in the previous section, all chemicals must be handled carefully with attention to personal safety. In addition, the use of laboratory chemicals, whether pure solids or solutions, requires care to be certain that little is wasted and that the purity of the substances and solutions is maintained.

Before you use any chemical, review the procedure to determine how much you will need. You should *never* draw material directly from a common vessel that others must also access. Instead, pour some of the solution or solid material into a clean dry vessel for use by yourself or your group. Then, all measurements of the amount of the material can be made from this portion. That way, if your technique contaminates the sample, only a small amount, not a whole bottle, is affected.

If you find that you have excess material after you have prepared your experiment, then treat this material as waste. You do not know if it has become contaminated in handling, so it should never be returned to the common stock.

B. Measurement by Mass

Measurement of the mass of a substance is a critical part of many experiments in general chemistry. It is essential that this be done in a precise and accurate manner. The *precision* is dictated by the instrument. We typically refer to a balance by the last digit that it provides on a gram (g) basis. Thus, a balance that can weigh a mass to a precision of 0.01 g is a centigram balance and one that weighs to the nearest 0.001 g is a milligram balance. The *accuracy* of a measurement depends on how the mass determination is done. Sloppy technique can mean large errors in measurement.

In measuring both mass and volume (next section), it is important to protect the original materials from contamination. Therefore, dispense only the amount you need to use; if you make a mistake and take too much, then discard the excess in a proper manner.

Electronic balances are very common in the undergraduate laboratory. These work by electronically detecting the effect of a weight placed on the pan. Milligram balances are so sensitive that they can detect changes in air currents. Therefore, each balance is equipped with a draft shield that has movable windows on the sides and on the top. If at all possible, these should be closed before recording data.

A balance must be zeroed before all measurements. To do this, close all the windows on the draft shield if present, then *gently* press the Tare button. After a moment, the display should respond with a reading of zero grams. Put your sample onto the balance and, after closing the windows on the draft shield if you can, record the mass. When you are finished, remove your sample, close all the windows, and rezero the balance.

It is very tempting to use the Tare button to make the mass of a weighing vessel or paper equal to zero grams in order to simplify calculations. In some cases this is justified, but it is a bad habit to use in a common balance room. The next time someone pushes the Tare button, *your information is permanently lost and you may not even know that the value of the Tare has changed.* Therefore, always zero the balance before weighing your vessel or paper. Then record the absolute weight (to the precision of the balance) of the vessel or paper. Subtract this mass from the mass of the vessel + sample to obtain the mass of the sample.

If milligram precision is required, then fingerprints and other soil can add detectable mass to an object. Therefore, between the initial and final weighing, all containers should be handled by tongs or, if necessary, protected by a piece of laboratory tissue.

In many cases, the most precise balances will be separated in a special area or room, where no transfers of chemicals should occur. In this case, you should go to the balance room to get an accurate measure of the weight of a container or weighing paper. Then, back in the laboratory, a chemical or solution is dispensed, often with a preliminary mass determination on a less precise balance. Finally, the container or paper plus the substance or solution is reweighed on the more precise balance.

There are many cases in the laboratory when you are instructed to use a mass of substance that is *close to* some particular value, for example, 2 g of metal salt. This means that, for the purpose of the experiment, there is some latitude in the mass of that substance. Any value within 10 to 20% of the indicated approximate mass is fine. However, in most cases it is important to know the precise mass of what you do use. So, even if there is latitude in the mass that you use, you must get a precise mass for your experiment. Thus, the notebook should record the mass determinations to within 0.001 or 0.01 g as indicated.

C. Measurement by Volume

Cleaning and Preparing Glassware

Many chemical procedures require the use of liquids, either pure substances or solutions. How much is used, and how precisely we determine the amount, varies widely from experiment to experiment. In some cases, when we say 10 mL of water, any value between about 5 and 15 mL is fine. Then a measurement based on markings on the side of a beaker or flask is adequate. In other cases, we will need to have exactly 10.00 mL of water, thus an instrument known as a volumetric pipet will be used.

Because solutions often contain reactive substances and because many solids dissolve readily in water, it is very important that the vessels used to handle liquids be very clean and, often, dry.

The first step in cleaning glassware is to remove any solids that are apparent on the glass. If you do not know what the solid is, then check with your instructor before cleaning the glassware. If the material is safe to dispose of in the drain, then a simple soap and water cleaning with tap water and a laboratory brush should be enough.

Rinsing with water or scrubbing with soap can leave behind a significant amount of other material. Therefore, before using most glassware it is essential that it be well rinsed with the purest water available in the lab. Usually, this will be deionized water. The rinsing does not require that the vessel be filled completely. Adding several small portions of deionized water, swirling to ensure that the water has wet the whole vessel, and then draining the vessel will be enough.

You can check for cleanliness by putting some deionized water in the vessel and then pouring the water out. Some water will stick to the inside of the vessel. This should appear as a clean sheet of liquid. If no beads form on the walls of the vessel, then it is clean and you may proceed. If beads form, wash the vessel with a soap solution and a brush, rinse with tap water, and rinse again with deionized water. Repeat until no water beads form on the inside.

In some cases, you will be dispensing a chemical or a solution directly into the vessel. In that case, it is often okay if you have a few drops of pure water present in the vessel. These will not affect the mass or the number of moles of added reactant.

Dry glassware is required when the added substance or solution has to remain dry or be undiluted. In those cases, drying the clean glassware means shaking or dabbing the last drops of water from the vessel and allowing the vessel to air dry for a few minutes. Some laboratories may have an organic solvent like acetone available. This can aid drying, but acetone can interfere with many procedures and it is flammable. Always check with your instructor regarding the use of acetone in drying glassware. Avoid drying using a stream of laboratory air. This air is *not* expected to be clean and dry and may contaminate your clean glassware.

Some experiments involve the use of a reactant or solvent that is of a known composition or concentration. To ensure that this is not changed, after the vessel is cleaned it must be *pre-rinsed* with the liquid that it will hold. To do this, add a few milliliters of the substance or solution to the vessel. Swirl to coat the entire inside of the vessel. Then discard the substance or solution in a proper manner. Do *not* return these rinses to the original container. Discard them as instructed.

Types of Volumetric Glassware

Two factors determine what glassware should be used to measure a liquid volume: convenience and precision. These are discussed in the chart that follows, which includes the major types of glassware found in general chemistry laboratories. All of these kinds of glassware are available in a variety of sizes. The size required is determined by the amount of liquid that needs to be measured and, for beakers and Erlenmeyer flasks, whether the glassware will be used for another purpose, such as carrying out a reaction.

Using Volumetric Glassware

Burets. Most burets used in general chemistry laboratories contain between 25 and 50 mL of solution. They are marked in 0.1-mL increments. But, as with any measured number, you can consistently get data to 0.05 or even 0.01 mL by looking very carefully at the level of the solution. The material dispensed from the buret must be the undiluted solution from the bottle you are using. To ensure this, the buret must be clean of all other material, including other solutions, and it must be pre-rinsed with the solution. The tip of the buret must be filled with the solution before the experiment is begun. Be very careful to check the tip for bubbles; these affect the accuracy of the measurement.

The flow of solution from a buret is controlled by a stopcock, which can be turned to permit the solution to leave the buret at a wide variety of rates. It is a good idea to practice with some water before using a valuable reagent. Try to get a smooth stream, a rapid flow of drops, and a flow of drops that is less than one per second. The very best control is to allow less than one full drop to form at the tip of the buret. Then, if appropriate for the experiment, a stream of deionized water from a wash bottle can be used to knock the drop off the buret tip and into the beaker or flask below.

Pipets. Two kinds of pipets are used for volumetric measurements. One, called a volumetric or transfer pipet, measures one particular volume to high accuracy. The other, a Mohr pipet, has gradations that permit different volumes to be dispensed very easily.

Both kinds of pipets are controlled in the same way, through a pipet bulb that is placed atop the pipet and used to pull solution into the pipet and, sometimes, to dispense the solution. There are several different kinds of bulbs. The simplest has a single opening that connects the bulb to the pipet. This is used to draw liquid into the pipet by squeezing the bulb and then placing it atop the pipet. The pipet is placed in the solution to be measured. The bulb is slowly released, and the suction draws the solution into the pipet. Be careful to keep the tip below the surface of the solution. If you do not, you may draw air very quickly into the pipet and cause the solution to shoot up into the bulb.

Type of glassware	Issues of convenience	Issues of precision and accuracy
Beakers	A wide opening at the top makes beakers easy to clean and to add material. A pouring lip makes dispensing easy.	The lines on the side of the beaker are rarely precise to more than 10%.
Erlenmeyer flasks	The narrow top is good for containing vapors and for pouring. The angle of the sides is useful for containing a stirred solution.	The lines on the side of the flask are rarely precise to better than 10%.
Burets	Burets have markings that allow for measurement of a range of volumes. The buret's stopcock can be used to dispense a steady stream of a solution or drops (even fractions of a drop).	Burets measure different volumes to a consistent precision (to 0.01 mL for a 25-mL buret).
Volumetric pipets	These allow the rapid measurement of particular volumes reproducibly and precisely.	The single-volume measurement of a volumetric pipet gives the greatest precision available (for example, 0.01 mL in 25 mL).
Mohr pipets	These allow the rapid measurement of a variety of volumes, with good precision.	The precision of Mohr pipets is usually greater than that of burets, sometimes 0.01 mL in a 5-mL measurement.
Droppers	Counting drops is a fast way to dispense small amounts of liquid.	The precision depends on a consistent drop size, and this requires a steady hand and constant angle. You should never assume two droppers have the same drop size.
Volumetric flasks	These provide an excellent way to take a measured amount of material and use it to prepare a solution of known concentration.	Volumetric flasks are the best way to prepare an exact amount of solution of a fixed concentration. They are generally precise to $\pm$ 0.01 mL. They come in many different sizes.

As the level of the solution nears the line you wish to reach, slow down on the rate by tightening the grip on the bulb. Draw a slight excess of solution into the pipet and then, without removing the bulb, lift the pipet out of the solution. With a simple bulb you should next smoothly but quickly slip a finger onto the top of the pipet in place of the bulb. Gently lifting this finger (a rolling motion may be best) allows air back into the pipet, and the solution will begin to flow out of the tip. Stop when the level of the liquid is back to the desired line.

If you have a volumetric pipet, you are ready to dispense the liquid directly from the volumetric pipet. Simply lift your finger and allow the solution to drain from the pipet. A small amount of liquid may remain in the tip. *The pipet is usually designed to have that liquid remain, so do not dispense it as part of the measurement.* A Mohr pipet works by measuring the difference between a starting and ending volume. Dispense the liquid by gently lifting your finger; stop the flow by putting your finger back on the pipet.

D. Techniques of Titration

No matter what kind of chemistry you do, even if you don't pursue a career as a chemist, you are likely to have to run a **titration.** A titration occurs when we add a chemical substance to a system gradually and observe the changes. In the laboratory, this usually means we use a solution with a known amount of reactant to determine the number of moles, and then the concentration, of the reactant in a second solution.

There are two reactants in any titration. The material added is called the *titrant* and the solution it goes into is the *titrand* (this latter term is rarely used). One solution contains a known chemical amount of a reactant. The other contains an unknown amount of a second reactant. The titrant is added to the titrand until some indication signals that the reactant in the titrand is all consumed. This is the *end point*. Good titration technique makes this end point very close to the volume at which a stoichiometric amount of the two reactants has been mixed and has reacted (the equivalence point).

A good titration reaction is quick and complete. The end point should be easy to detect ("sharp"). In practice, end points are easy to overshoot. In those cases, the end point is some unknown amount beyond the equivalence point. To get a more accurate determination of an end point close to the equivalence point, good technique demands that you do more than one trial of any titration. The first trial may be off, but it gives a much better idea of where the end point will be, so subsequent trials can be much more accurate. This additional accuracy comes from making the approach to the end point as gradual as possible.

There are three common types of titration reactions. One type involves acid–base reactions, which are important in determining concentrations of acids and bases. A second type involves oxidation–reduction reactions. The third type of titration is the compleximetric titration.

Dispensing the Titrant

When dispensing the titrant, it is okay to add a solution as a stream *only* if you are certain that you are not near the end point. As you approach the end point, start adding titrant a drop at a time. Often you know the end point is near because you can see, temporarily, the color change that will signal the true end point. Be careful to mix the solution completely at the first indication of color change. You have reached the end point of the titration when the color change persists in a well-mixed solution for at least 30 s. The last of the titrant should be added as fractions of a drop. You do this by allowing a drop to start to form at the tip of the buret, then use a gentle stream of deionized water from a wash bottle to rinse the drop into the solution below.

The end point may be ambiguous in some cases. Therefore, record the volume of the buret and the appearance of the titration solution at several points. You may then examine the data to find the place that you think is closest to the true end point. If you overshoot, of course, you will lose information that may be important. So, *go slowly*.

Acid–Base Titrations

Most acid–base titrations involve a solution of an acid and a solution of a base. In the simplest case, the acid and the base react in a 1:1 ratio to give the corresponding conjugate base and conjugate acid:

$$\text{acid}_1 + \text{base}_2 \rightarrow \text{base}_1 + \text{acid}_2$$

This represents the Brønsted–Lowry theory of acids and bases, and is one of the clearest ways to describe how acids and bases react with each other in solution.

In all practical titrations in aqueous solution, either the starting acid or the starting base is strong. Thus, one of the reactants is either H_3O^+ or OH^-, giving water as a product:

$$H_3O^+ + B \rightarrow B\text{—}H^+ + H_2O$$
$$HA + OH^- \rightarrow A^- + H_2O$$

How do we know that an acid–base titration is complete? The simplest way is by means of a colored acid–base indicator. These indicators are substances that, in the presence of even a small amount of excess acid or base, rapidly change color. Therefore, we can add acid or base until the indicator changes color and then we know we've completed the titration. *For monoprotic acids and bases*, at completion, we have added an amount of acid (or base) exactly equal to the **number of moles** of the base (or acid) in the original solution. From this number of moles and the volume of the solutions we can determine the molarity of the unknown solution. The fundamental equation is

$$\text{moles acid} = \text{moles base}$$
$$c_{acid}V_{acid} = c_{base}V_{base}$$
$$c_aV_a = c_bV_b$$

We can solve for any of the four quantities in this equation if we know the other three. We can also work with a known *mass* of an acid or a base and use the molar mass to find moles acid or base:

Known mass of acid with unknown [base]	**Known mass of base with unknown [acid]**
moles acid = moles base	moles acid = moles base
$\dfrac{\text{mass acid}}{\text{molar mass of acid}} = c_bV_b$	$c_aV_a = \dfrac{\text{mass base}}{\text{molar mass of base}}$

When the acid and the base do not react in a 1:1 mole ratio, you must use the stoichiometry of the balanced equation to find the relationship between moles acid and moles base.

Standards for Acid–Base Titrations

Two kinds of standards are used in analytical chemistry. The first type, called a *primary* standard, is a highly purified compound that is weighed accurately on a balance before it is dissolved and used in the titration. A *secondary* standard is one whose concentration is determined from a primary standard and then used in other reactions. Secondary standards are usually less expensive and easier to handle than primary standards. However, most secondary standards will change in concentration with time. For example, strong bases will react with carbon dioxide in the air.

E. Dilution Calculations

In many experiments and procedures, volumetric glassware is used to prepare solutions from more concentrated solutions known as stock solutions. Here is a summary of the concepts and steps involved.

When we dilute a solution, we increase the volume without changing the chemical amount (commonly, moles) of the substances that are dissolved in the solution. This means that, although the volume V and the concentration c change, the number of moles does not. If we express concentration in units of amount per volume, then concentration times volume, $c \times V$, is equal to the amount. If we are tracking moles,

$$n_{\text{initial}} = n_{\text{final}}$$
$$c_{\text{initial}}V_{\text{initial}} = c_{\text{final}}V_{\text{final}}$$

It is common to abbreviate final as f and initial as i, so this equation becomes $c_iV_i = c_fV_f$. Let's use this in some sample problems

Dilution of a Known Concentration by a Known Amount

If we dilute 2.50 mL of 0.100 M Cu^{2+} to 25.00 mL in a volumetric flask, what is the final concentration of Cu(II) ion?

Answer. In this case, c_i is 0.100 mol L^{-1}, V_i is 2.50 mL, and V_f is 25.00 mL. We solve for $c_f = c_iV_i/V_f = (0.100 \text{ M})(2.50 \text{ mL}/25.00 \text{ mL}) = 0.0100$ M. Note that we do not have to change the volume to liters, because both volumes are given in the same units, milliliters. The V_i/V_f ratio is a constant value as long as the volume units are the same. Here,

$$\frac{V_i}{V_f} = \frac{2.50 \text{ mL}}{25.00 \text{ mL}} = \frac{0.00250 \text{ L}}{0.02500 \text{ L}} = 0.100$$

Determining How Much to Dilute a Solution

We need to prepare 0.500 L of a solution of 0.0250 M Fe^{3+}. We have a stock solution of 0.528 M Fe^{3+}. How should we prepare the solution?

Answer. Here we are seeking the value for V_i. This will be

$$V_i = (c_f/c_i)V_f = (0.0250 \text{ M}/0.528 \text{ M})\ (0.500 \text{ L}) = 0.0237 \text{ L}$$

In this case, we will take 0.0237 L, or 23.7 mL, of the stock solution and place it in a 0.500-L volumetric flask. Adding water, then mixing, to give a 0.500-L total will give the required solution. Note that it is incorrect to say we will add 472.3 mL of water, because that assumes that volumes are additive. This is an assumption we must make sometimes, but never if we can use volumetric glassware with precise volume markings.

Finding a Concentration When Solutions Are Mixed

There are times that the most efficient, though not the most precise, way to make a solution is by mixing simple volumes. For example, if we mix 3.80 mL of 0.50 M NH_3, 0.30 mL of 0.100 M Ag^+, and 0.90 mL of water, what is the concentration of the $[Ag(NH_3)_2^+]$ that will form, assuming that volumes are additive and that all of the silver ion becomes $[Ag(NH_3)_2^+]$?

Answer. The total volume of this mixture will be, approximately, 5.00 mL. We note that the number of moles of $[Ag(NH_3)_2^+]$ will equal the number of moles of Ag^+, so we can calculate the concentration of $[Ag(NH_3)_2^+]$ from the concentration of Ag^+ added:

$$c_f = c_iV_i/V_f = (0.100 \text{ M})(0.30 \text{ mL}/5.00 \text{ mL}) = 0.0060 \text{ M}$$

IV. Principles of Spectrophotometric Analysis

A. The Measurement of Color

The human eye and brain are extremely sensitive to color. When the eye and brain sense a spectrum of visible light that is the same as that of the sun, then they perceive either white or a shade of gray. If the eye and brain detect only a section of the spectrum, then a color is perceived.

There are two ways an object can come to have color. First, the object can be a source of light of a restricted wavelength. Under these circumstances, the observer perceives the wavelength that comes from the emission of light by the object. Think of the bright yellow lights that are often used for exterior lighting. These use the emission of light by sodium vapor to give a bright light that is narrowly restricted to the region at about 590 nm (one nanometer is 1×10^{-9} m), at the center of the yellow region of the spectrum.*

Color can also arise when an object absorbs some wavelengths but reflects or transmits others. This selectivity arises from the absorption of light by the object. An observer perceives the *complement* of the light that is absorbed. A color wheel (Figure IN-1) helps us to understand and predict the appearance of simple absorption. For example, when light of blue wavelengths is absorbed, then light of violet, green, yellow, orange, and red wavelengths is transmitted or reflected. This combination appears to our eyes as orange.

The eye and brain have limitations in ascertaining which wavelengths of light are absorbed by a sample. For example, say you see the color red. This can arise because only green light is being absorbed by a sample, in which case your brain combines the perception of transmitted red, orange, yellow, blue, and violet light as red. On the other hand, the color red is perceived in cases where *all* wavelengths of light *except* red are absorbed and red light wavelengths are reflected or transmitted. Your senses cannot tell the difference.

Another characteristic of the eye and brain is that they are more sensitive to certain wavelengths and less sensitive to others. The peak sensitivity is in the region where the sunlight is concentrated: around 500 nm. This is the green–yellow region

Figure IN-1 A color wheel to determine complementary colors

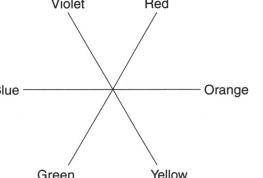

*You can tell that this light is of only a narrow wavelength range because, if you look at a color picture illuminated by such a sodium vapor light, the picture appears sepia, regardless of the component colors.

of the spectrum. The great sensitivity of the human visual system to green–yellow light is the reason this color is often used on emergency vehicles and taxicabs.

Scientists study light and color quantitatively through the use of spectrophotometry—the quantitative determination of light intensity by wavelength. The signal from an electronic detector changes depending on the intensity of light striking the detector. These detectors are (generally) equally sensitive over a wide range of light wavelengths. To determine the amount of light at different wavelengths, one must separate the light into its components. The standard way to do this is through the use of a prism or diffraction grating that bends beams of different wavelengths by different amounts. The gathering of information concerning intensity versus wavelength is called *spectrophotometry*, and the corresponding instruments are called *spectrophotometers*.

Spectrophotometry can be performed across the electromagnetic spectrum (Figure IN-2), from gamma rays (useful in nuclear chemistry) to radio waves (used in astronomy). Spectrophotometers designed for wavelengths in the ultraviolet (180–380 nm) and visible (380–700 nm) regions are common. They are referred to as UV-visible spectrophotometers.

B. Absorbance and Transmittance

There are several different ways of quantifying the amount of light that comes through a sample. The simplest to understand is percent transmission (%T), which is the percentage of light that comes through the sample at a given wavelength. A far more meaningful number in practice is the absorbance, A. This is equal to a logarithmic function of %T:

$$A = -\log_{10}\left(\frac{\%T}{100}\right)$$

Absorbance is an open-ended scale and has no units. When all of the light gets through (%T = 100), then A = 0. When %T = 10%, A = 1.0. When %T = 1%, then A = 2.0. When no light is transmitted (%T = 0), then A = ∞.

The relationship between absorbance and concentration can be a simple one:

$$A = \varepsilon c l$$

where *c* equals concentration in moles liter^{-1} (mol L^{-1}), *l* equals path length (typically 1 cm), and ε (this is the lowercase Greek letter epsilon) equals the *extinction coefficient*, which in these systems has units of L mol^{-1} cm^{-1}.

Figure IN-2 The electromagnetic spectrum

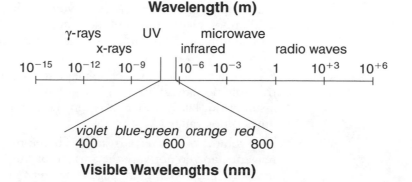

The linear relationship between absorbance and concentration is known as *Beer's law*, and it is the basis for a great deal of spectrophotometric analysis. Beer's law is reliable only when A is small—say under 2—and it relies on the assumption that the nature of the substance under study does not change with concentration.

The extinction coefficient indicates how much of the light that strikes a sample is absorbed by a given concentration of the chemical substance. Small extinction coefficients are in the range 1 to 1000 L mol^{-1} cm^{-1}. Large extinction coefficients can range up to 100,000 or more. You can also think of the extinction coefficient as an efficiency indicator, and indeed systems that need to operate at high efficiency (such as the photosynthetic apparatus in plants) have very large extinction coefficients. Food colorings, which give a lot of color for very little substance, also have large extinction coefficients.

C. Deviations from Beer's Law

Beer's law suggests that a linear relationship exists between absorbance and concentration. This relationship holds for many, but not all, substances. Deviations can occur for at least three reasons:

1. The molecules that absorb light interact with one another instead of absorbing light independently.
2. The light-absorbing substance forms incompletely from the added reagents because of equilibrium or side reactions.
3. The light-absorbing substance forms slowly compared to the time required to make an absorbance measurement.

In all three cases, we can find out whether Beer's law applies by testing a series of different solutions prepared with different measured amounts of reagents. This is used to construct a graph where the y-axis has units of absorbance and the x-axis has units of concentration. If the graph of A versus c is a straight line, then Beer's law applies. If it is not a straight line, then we can still determine c for an unknown solution by comparing A with the curve that the standard solutions provide.

D. Spectrophotometric Samples

Most spectrophotometers use small glass or plastic cuvettes to hold the sample. They have a square design, with two parallel clear sides for the light to shine through. A 1-cm cuvette typically holds 3 mL of solution.

The most accurate technique requires precise control of solution concentrations, so volumetric glassware, including volumetric flasks, is used to prepare samples. Then the sample is transferred to the cuvette to fill it.

It is also possible to get good results by preparing the test sample right in the cuvette, saving glassware and time. If this is done, you must remember three points:

1. All samples must have the same volume. If you mix 2.00 mL of a reagent with 0.30 mL of a second reagent, don't forget to add enough of the correct material—for example, water or buffer—to make the volume the same total volume in each case.
2. Preparing a solution by mixing measured amounts of other solutions assumes that the volumes are additive. In these labs, we have verified that it is a reasonable assumption, but in other situations be aware that the additivity of solution volumes is not valid.
3. The volume of a cuvette is small. Therefore, plan to work with small amounts of reagents, and do not vary the amount of the "main" reagent. This can be done by using 2.0 mL of the main reagent in every spectrophotometric sample.

Systematic errors can arise in preparing spectrophotometric samples. All glassware should be clean and dry. In addition, cuvettes are easily dirtied or scratched, and these flaws can absorb light. Finally, one or two sides of the cuvette may be frosted. Do not measure a sample unless the light will pass through clear, clean surfaces.

E. Using the Spectronic-20

Many different spectrophotometers are in use today. Ask your instructor which type you will use. Here we give directions for the most common instrument.

The Spectronic-20 spectrophotometer (Figure IN-3) uses a tungsten lamp as its source of light for the analysis of compounds that absorb in the UV-visible range ($\lambda = 350-800$ nm). The optical components of the instrument itself focus the incoming light with mirrors. The user then selects the particular wavelength of light that finally passes through the sample. Particular wavelengths can be selected by means of an outside knob that allows you to "dial in" the wavelength you desire. The selected wavelengths of light reflect off the grating and are focused by a second mirror to pass through a sample. Light emerging from the sample strikes a detector that analyzes its intensity. A comparison of the amount of light entering the sample with the amount of light leaving the sample gives a measurement of the amount of light that was absorbed by the sample.

Several techniques must be understood and mastered before you attempt to use any spectrophotometer. Good results depend on a measurement of the light that is absorbed by a solution of the sample. The lid of the sample compartment must be closed before taking a measurement so that all of the selected light passes through the sample. The instrument must be adjusted to correctly register both zero and 100% transmission before each use. The procedure outlined here is a good one to follow.

1. Allow about 20 min for the instrument to warm up.
2. Adjust the wavelength control knob to the desired wavelength.
3. Prepare the instrument to make accurate measurements for your particular solution by making the following adjustments:
   (a) With no sample in the sample compartment, adjust the dark current control to read 0% transmission (%T).
   (b) With a blank solution in the sample compartment, adjust the light control knob to read 100% transmission. The blank solution may be distilled water, or it may be the solution that will eventually contain your sample. Specific directions for making up a blank solution are generally given within a laboratory procedure.

Figure IN-3 The controls of the Spectronic-20 spectrophotometer

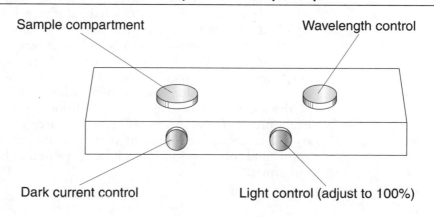

Sample compartment   Wavelength control

Dark current control   Light control (adjust to 100%)

4. With these steps done, you are ready to measure the absorbance of your sample. Insert the cuvette containing your sample into the sample compartment, close the lid, and record both the absorbance and the % transmission ($\%T$) readings. The scale for $\%T$ may be easier to read accurately, and it can readily be converted into absorbance through the relationship $A = -\log(\%T/100)$.

F. Calibration Curves

Many spectrophotometric procedures require you to prepare a calibration curve for analysis of unknowns. A calibration curve is *your* ruler to measure a chemical system. This must be done well, because all later data will be referenced to this one set of measurements.

You start the process of calibration with an accurately determined concentration of a substance in a stock solution. This is then diluted to give a set of solutions that span the range of possible concentrations in the experiment. Sometimes these standards are placed right in cuvettes. If necessary, these standard samples are combined with other reagents to give the sample for the instrument.

In Figure IN-4, we present data for the calibration curve we might use in determining the concentration of the dye carmine indigo. It does not need a second reagent to give a suitable spectrophotometric sample. We take a sample of a solution of carmine indigo and place it in a cuvette with no further manipulation. The range of possible values we must "cover" with the curve is equal to the range of values we may have in the actual sample.

The figure shows the data in three ways. First, notebook entries indicate the volume of stock solution mixed with water to prepare the standard solution. Note that the notebook indicates exactly how the sample was prepared. Second, the solutions were transferred to a spectrophotometer and the absorption spectrum was measured, giving A values for our notebook (again, record *all* essential data in the notebook). The acceptable range of values of A for spectrophotometry is generally 0.05–1.5. We can see that the data are in the acceptable range of values, so we can use this method to prepare a calibration curve.

These data were transformed into the graph at the bottom of the figure. Note that the graph has five data points, one for each standard, and a line connecting them. Dotted lines are used to extend the curve to the end of the graph region. These are *extrapolated* lines, because we did not actually measure these regions. One should only use an extrapolation of a graph in emergencies.

The data analysis suggested in the figure should be done *during the lab period*, including a preliminary calibration curve. If any data points seem to lie off the curve of the other data, then there is ample time to remeasure those particular points.

In the second kind of calibration curve, we mix solutions of a substance with another reagent, yielding a colored substance for measurement. We prepare the standards, as before, to cover the range of possible concentrations in our unknown solution. All of these—standards and unknowns—are then mixed to prepare solutions for the spectrophotometer. This is often done right in the cuvette, although very accurate work requires that this step be done in a volumetric flask. We must mix a consistent volume of our standards and unknowns with a consistent amount of the solutions that contain the additional reagent or reagents. These are then examined with the spectrophotometer. After verifying with a good spectrophotometric calibration experiment that the data are consistent, we plot a calibration curve, and this allows us to read the curve to find how much of a substance is in the unknown.

Figure IN-4 Schematic preparation of a calibration curve

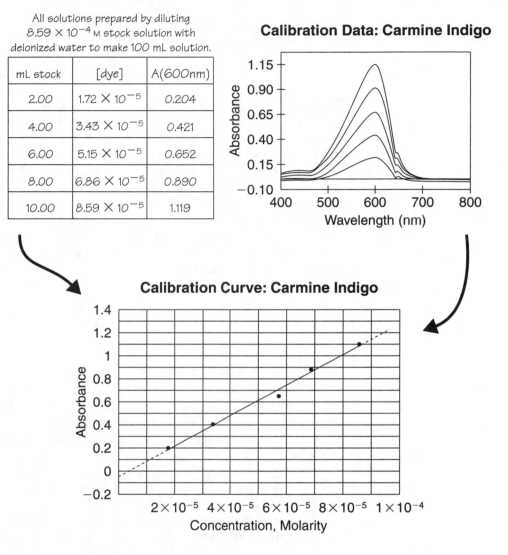

Notebook records

All solutions prepared by diluting 8.59×10^{-4} M stock solution with deionized water to make 100 mL solution.

mL stock	[dye]	A(600nm)
2.00	1.72×10^{-5}	0.204
4.00	3.43×10^{-5}	0.421
6.00	5.15×10^{-5}	0.652
8.00	6.86×10^{-5}	0.890
10.00	8.59×10^{-5}	1.119

Calibration Data: Carmine Indigo

Calibration Curve: Carmine Indigo

V. Quantitative Interpretation of Chemical Measurements

A. Chemical Reaction Stoichiometry

The measurement of the amounts of substances involved in a chemical reaction is known as *chemical stoichiometry*. It is an essential part of most general chemistry courses, and here we review only a few major points.

Measurement of Chemical Amount

Chemists cannot measure the number of molecules in a sample directly, except under very unusual circumstances. Instead, they measure the amount by other means, especially by measuring mass. We connect a count of molecules or formula units with mass through the formula mass, which is determined by multiplying the atomic mass of each element present by the number of times its symbol appears in the formula.

Atomic and formula mass are linked to measurable masses in the laboratory. For chemists, this mass is the *gram*. We define the mass of one mole of a substance as equal to the formula mass, expressed in grams. Molar mass is the mass, in grams, of one mole of a substance.

For example, one mole of carbon dioxide, CO_2, has a mass of 12.011 g + 2(15.9994 g) = 44.0098 g (calculator answer) or 44.010 g (rounded answer):

$$\text{molar mass } CO_2 = 44.010 \text{ g mol}^{-1}$$

The molar mass of a chemical substance is one of its most important properties. Why? Because we cannot directly measure the number of formula units or molecules in a sample; there are no "molemeters" in the laboratory. Instead we can easily measure the mass of a sample. And from the mass of a substance and its molar mass, we can determine the number of moles in a sample. The molar mass is the constant of proportionality relating mass and moles.

Molar mass can also be used to tell us how many grams of a substance give a certain number of moles. This is critical if chemists are to measure the right amount of substance. For example, if we want to know the number of moles of iron in a sample of iron that has a mass of 223.22 g, we can use the molar mass of iron to carry out a unit conversion calculation:

$$223.22 \text{ g Fe} \times \frac{1 \text{ mol Fe}}{55.845 \text{ g Fe}} = 3.9971 \text{ mol Fe}$$

Chemical Amounts in Reactions

Chemical substances can react with one another to form new substances. This is called a *chemical reaction* and is shown in symbols by a chemical equation. In a chemical equation, the substances that react (called *reactants*) appear to the left of an arrow and the new substances that are produced (called *products*) appear to the right of the arrow.

$$\text{reactants} \rightarrow \text{products}$$

In addition to showing reactant and product substances, a correctly written chemical equation shows the ratio these substances have to one another. Not all substances react in a one-to-one ratio. The ratio of substances in a chemical equation is given by *coefficients*, the numbers written in front of the chemical formula representing the substance. For example, the chemical equation for the reaction of hydrogen gas and nitrogen gas to form ammonia gas is,

$$3 \text{ H}_2 + \text{N}_2 \rightarrow 2 \text{ NH}_3$$

The coefficients in a balanced equation show how many molecules react and are produced in the chemical reaction. Thus, this equation means "1 molecule of N_2 + 3 molecules of H_2 react to form 2 molecules of NH_3." Because the mole is a group counting number that can be used to count large numbers of atoms and molecules, we can also say the equation means "1 mole of N_2 + 3 moles of H_2 react to form 2 moles of NH_3." We can then use this to determine how many moles of ammonia will be produced from the reaction of 0.50 mole of hydrogen and an excess of nitrogen according to the balanced equation. We can use the ratio

$$\frac{2 \text{ mol NH}_3}{3 \text{ mol H}_2}$$

as a conversion factor between moles of H_2 and moles of NH_3:

$$0.50 \text{ mol } H_2 \times \frac{2 \text{ mol } NH_3}{3 \text{ mol } H_2} = 0.33 \text{ mol } NH_3$$

Many of the calculations in the laboratory actually require that we begin with the mass of one substance and end with the calculation of the mass of another substance. This means we must add a mass-to-mole calculation to the beginning of the process and a mole-to-mass calculation to the end.

For example, iron metal reacts with oxygen (is oxidized) to form iron(III) oxide as represented by the equation $4 \text{ Fe } (s) + 3 \text{ O}_2 (g) \rightarrow 2 \text{ Fe}_2O_3 (s)$. We can use stoichiometry to determine how many grams of iron are needed to produce 100.0 g of iron(III) oxide. It is helpful to start with a "calculation map":

$$100.0 \text{ grams } Fe_2O_3 \rightarrow \text{moles } Fe_2O_3 \rightarrow \text{moles Fe} \rightarrow \text{grams Fe}$$

$$100.0 \text{ g } Fe_2O_3 \times \frac{1 \text{ mol } Fe_2O_3}{159.691 \text{ g}} \times \frac{4 \text{ mol Fe}}{2 \text{ mol } Fe_2O_3} \times \frac{55.845 \text{ g Fe}}{1 \text{ mol Fe}} = 69.9413 \approx 69.94 \text{ g Fe}$$

Limiting Reactants

The coefficients of a balanced equation are the exact number of moles of reactants needed as well as the exact number of moles of products formed. The phrase *stoichiometric amounts* describes the case when the relative amounts of the reactants match the amounts required by the balanced equation. When you do *not* have stoichiometric amounts of reactants, one of the reactants will be used up *before* the others. This will stop the reaction. The reactant that is used up is called the **limiting reactant** or limiting reagent (L.R.) because it is the substance that "limits" the amount of product that can be formed.

Determination of a limiting reactant in a chemical reaction is not always so obvious. One method used to solve limiting reactant problems is to determine the amount of product that will result using *each of the given quantities one at a time. The reactant that produces the smallest amount of the same product is the limiting reactant.* When the L.R. is gone, the reaction stops. All of the limiting reactant is gone, but some of the other reactant(s) remain. These are referred to as **excess reactants.** Only the limiting reactant matters in determining the expected amount of product.

For example, copper(II) sulfate and barium chloride solutions react to produce a precipitate of barium sulfate. (A precipitate is a solid that appears "suddenly" or "precipitously" when two solutions are mixed.) The reaction occurs according to the equation

$$CuSO_4 (aq) + BaCl_2 (aq) \rightarrow CuCl_2 (aq) + BaSO_4 (s)$$

If we have 5.255 g of copper(II) sulfate and 8.230 g of barium chloride reacting, then to calculate the expected mass of $BaSO_4$, we need to do two calculations. The first calculation determines the mass of barium sulfate produced when you start with 5.255 g of copper(II) sulfate. This calculation assumes that the copper(II) sulfate is completely consumed. The second calculation determines the mass of barium sulfate produced when you start with 8.230 g of barium chloride, thus assuming that the $BaCl_2$ is completely consumed. The limiting reactant is the one involved in the calculation resulting in the smaller amount of product. The smaller amount of product is the only correct answer.

$$5.255 \text{ g CuSO}_4 \times \frac{1 \text{ mol CuSO}_4}{159.608 \text{ g CuSO}_4} \times \frac{1 \text{ mol BaSO}_4}{1 \text{ mol CuSO}_4}$$

$$\times \frac{233.392 \text{ g BaSO}_4}{1 \text{ mol BaSO}_4} = 7.6843 \text{ g BaSO}_4$$

$$8.230 \text{ g BaCl}_2 \times \frac{1 \text{ mol BaCl}_2}{208.236 \text{ g BaCl}_2} \times \frac{1 \text{ mol BaSO}_4}{1 \text{ mol BaCl}_2}$$

$$\times \frac{233.392 \text{ g BaSO}_4}{1 \text{ mol BaSO}_4} = 9.2242 \text{ g BaSO}_4$$

The first calculation produces the smaller amount of $BaSO_4$. This means that $CuSO_4$ is the limiting reactant. The amount of $CuSO_4$ present at the beginning of the reaction limits the amount of product to only 7.684 g $BaSO_4$. The expected mass of precipitate is, therefore, 7.684 g $BaSO_4$. The other reactant, $BaCl_2$, is in excess.

Reaction Yield

The amount of product actually obtained in a reaction is called the *actual yield* of the reaction. The amount of product predicted by stoichiometry is called the *theoretical yield*. The actual yield should never be greater than the theoretical yield, but it can be less—much less in some cases. The theoretical yield and the actual yield form a ratio referred to as the *percentage yield*, where

$$\text{percentage yield} = \frac{\text{actual yield}}{\text{theoretical yield}} \times 100$$

As an example, the synthesis of silver chloride involves mixing known amounts of a chloride salt—say, potassium chloride—and silver nitrate, a soluble silver salt:

$$KCl \ (aq) + AgNO_3 \ (aq) \rightarrow AgCl \ (s) + KCl \ (s)$$

If we mix one mole of KCl and one mole of $AgNO_3$, we should get one mole of AgCl. This would have a mass of 143.32 g of AgCl. This mass is the theoretical yield. In a typical reaction, though, only 125 g of AgCl may be isolated. This is the actual yield. The percentage yield is then

$$\frac{125 \text{ g AgCl}}{143.22 \text{ g AgCl}} \times 100 = 87.3\%$$

B. Uncertainty Analysis

Chemistry experiments usually require measurement of one or more parameters that describe the particular chemical system studied. Most laboratory measurements are very common ones such as mass, temperature, volume, or pH. It is important that you assess the reliability or trustworthiness of your measurements before you draw conclusions based on your experimental data.

Experimental results are judged using two criteria: accuracy and precision. The **accuracy** of a measurement tells us how closely it agrees with the true value of that parameter. Chemical properties such as density have known values. For these properties, the difference between our measured value and the "true" value is an indication of the measurement's accuracy. The **precision** of the measurement indicates how closely each measurement agrees with the other measurements taken for the same parameter.

Accuracy in Measurement

Several types of error affect your results: systematic error, random error, and personal error. **Personal errors** such as spillage and careless measurements must be noted whenever they occur. Personal errors affect both the accuracy and the precision of your results. The most common kind of personal error occurs when something is done too quickly. For all measurements, take your time and, if possible, repeat the measurement. When transferring material, be sure that you transfer everything that can be properly delivered by a measuring device. With practice, such errors will decrease and not greatly affect your experimental results.

Systematic errors occur when there are flaws in the measuring device, which result in measurements that always occur in the same direction (too high or too low). Systematic errors affect the accuracy of your result. An example of systematic error is an improperly calibrated thermometer or balance. When we are aware of a systematic error, we try to correct it. In general, however, we are usually unaware of the error built into a particular measuring device. There are occasions when we seek the *difference* between two measurements (mass gained or lost, temperature change, etc.). In those cases, systematic errors *may* be equal to each other and can be ignored.

Random errors occur because of small, uncontrollable variables in one's ability to obtain exactly the same numerical value for numerous measurements of the same quantity. Random errors can be decreased but not entirely eliminated. Successive measurements of the same property result in approximately but not exactly the same numerical value. Measurements cluster about an **average value** or **mean,** with some measurements slightly higher and some slightly lower than the mean value. The difference between a particular measurement and the mean value is called its *deviation* from the mean. The degree of deviation from the mean is used to assess the precision of experimental results. Slight deviations from the mean represent good experimental precision; large deviations indicate poor precision.

Precision in Measurement

The precision of experimental results is calculated using the average deviation from the mean value. The mean is found by summing the individual measurements (X_i) and dividing by the total number of measurements summed (N).

$$\text{mean} = \overline{X} = \frac{\sum_{i=1}^{N} X_i}{N}$$

When only two or three results are available, we must rely on the range of values from lowest to highest as an indication of experimental precision. For example, suppose that you determine the density of an unknown metal as 2.978, 2.885, and 2.994 g/cm^3. The average value is

$$\frac{2.978 + 2.885 + 2.994}{3} = 2.946 \text{ g/cm}^3$$

The precision of these measurements is shown in the range of values reported; from the lowest, 2.885 g/cm^3, to the highest, 2.994 g/cm^3, there is a difference of 2.994 − 2.885 = 0.109 g/cm^3. Thus, all reported values fall within an average deviation of 0.109/2 = ±0.054 g/cm^3 from the mean and the results are reported as 2.946 ± 0.054 g/cm^3. The ± notation indicates the precision of the experimental results as it tells how closely the values are clustered (within the bounds of 0.054 higher and lower than 2.946).

Statistical analysis is used to determine the precision of experimental results for larger bodies of data, ideally 20 or more measurements, although it is often used whenever $N > 3$. A statistical treatment of random error gives the standard deviation as a measure of experimental precision. The **standard deviation** (s.d.) is the square root of the sum of the squares of each measurement's deviation from the mean divided by $(N - 1)$.

$$\text{s.d.} = \sqrt{\frac{\sum\limits_{i=1}^{N} (X_i - \overline{X})^2}{N - 1}}$$

Many calculators and spreadsheet programs include standard deviation functions. To calculate the standard deviation, find the mean value $\overline{X}$, find each deviation from the mean $(X_i - \overline{X})$, square each deviation, sum the squares, divide by $N - 1$, and finally take the square root of that number.

Example. Find the standard deviation for these experimental results: 1.1244, 1.1198, 1.1232, 1.1190, 1.1204 g/L.

Measurements, g/L	$(X_i - \overline{X})$	$(X_i - \overline{X})^2$
1.1244	0.00304	0.00000924
1.1198	−0.00156	0.00000243
1.1232	0.00184	0.00000338
1.1190	−0.00236	0.00000557
1.1204	−0.00096	0.000000922

$\overline{X} = 1.1213_6{}^*$

$$\sum_{i=1}^{N} (X = \overline{X})^2 = 0.0000215$$

$$\frac{\sum\limits_{i=1}^{N} (X - \overline{X})^2}{N - 1} = 0.00005375$$

$$\text{s.d.} = \sqrt{\frac{\sum\limits_{i=1}^{N} (X - \overline{X})^2}{N - 1}} = 0.0023_2$$

*The subscripted digit is the first insignificant digit. It is carried through the calculation and rounded off at the end.

The standard deviation is 0.0023_2 and is rounded to the same number of decimal places as the mean. The result is reported as mean ± s.d.; 1.1214 ± 0.0023 g/L.

Uncertainty Analysis and Propagation of Error

There is some uncertainty in all measurements we do in the laboratory. **Uncertainty analysis** estimates the error introduced into a measurement due to random error. The instruments and glassware we use in the lab vary in their abilities to give precise measurements. The range of their precision is indicated by a ± symbol followed by numbers that indicate the degree of precision for that measurement. For example, a mass of 50.00 ± 0.01 g indicates a balance that is precise to ±0.01 g in any mass measurement; the "actual" mass lies between 50.00 + 0.01 g and 50.00 − 0.01 g. We can be certain only that the substance has a mass somewhere between 49.99 and

50.01 g. A mass of 50.000 ± 0.001 g indicates a balance precise to ±0.001 g. The second balance is more precise than the first balance because it results in masses precise to thousandths rather than hundredths of a gram.

All directly measured parameters in an experiment have uncertainties. When these measurements are used to calculate some final result, their individual uncertainties will contribute to an overall uncertainty in the final calculated result. *Uncertainty is propagated in a calculation using directly measured parameters that contain uncertainties.* The final, overall uncertainty in a calculated result is obtained by doing three calculations. Using the initial uncertainties, we calculate the lowest possible value, the central value, and the highest possible value. Rules for these calculations follow.

- The symbol ± gives a high and a low limit for each measurement. This results in three values for each measurement: a low limit value, a central value (the measurement itself), and a high limit value.
- For addition and multiplication, add or multiply limits on the same side (high limit with high limit; low limit with low limit).
- For subtraction and division, subtract or divide limits on the opposite side. In subtraction problems, the high limit results when the minuend (top number in a subtraction problem) is the high limit value for that measurement and the subtrahend (bottom number in problem) is the low limit value for the measurement. In division problems, the high limit results when the numerator is the high limit value and the denominator is the low limit value for the appropriate measurements.

As an example, suppose you perform an experiment to measure the density of an unknown solid substance by water displacement. You measure the mass of your sample as 8.846 g on a balance that has an uncertainty of ±0.001 g.

	Low-side limit	Central value	High-side limit
Mass of solid (g)	8.845	8.846	8.847

The volume of the solid is the difference in water volumes (with and without the solid) using a 100-mL graduated cylinder that has an uncertainty of ±0.2 mL.

	Low-side limit	Central value	High-side limit
Volume of water (mL)	24.6	24.8	25.0
Volume of water + solid (mL)	27.4	27.6	27.8

To obtain density, we must first determine the volume of water displaced, then divide mass by this volume. The volume of the solid is determined by subtraction:

(volume of water + solid) − (volume of water) = volume of solid

Using the rules for subtraction, opposite side limits are used. This results in the following volumes.

Low-side limit: $27.4 - 25.0 = 2.4$ mL
Middle value: $27.6 - 24.8 = 2.8$ mL
High-side limit: $27.8 - 24.6 = 3.2$ mL

The three sets of values needed for a calculation of density are

	Low-Side Limit	Central Value	High-Side Limit
Mass of solid (g)	8.845	8.846	8.847
Volume of solid (mL)	2.4	2.8	3.2

The density calculation requires division, so again opposite sides are combined to get the high- and the low-side limits for the density.

Low-side limit: $\dfrac{8.845 \text{ g}}{3.2 \text{ mL}} = 2.76 \approx 2.8$ g/mL

Middle value: $\dfrac{8.846 \text{ g}}{2.8 \text{ mL}} = 3.159 \approx 3.2$ g/mL

High-side limit: $\dfrac{8.847 \text{ g}}{2.4 \text{ mL}} = 3.68 \approx 3.7$ g/mL

The high and low limits give the uncertainty in the end result: density = 3.2 ± 0.4 g/mL. The calculated density value contains approximately 10% uncertainty as a result of the initial instrumental uncertainties carried or propagated through the subtraction and division processes. The final percentage rounds to one significant figure in this case.

$$\frac{0.4}{3.2} \times 100 = 12.5 \approx 10\%$$

C. Graphing Laboratory Data

Principles of Good Graphing

The results of many experiments in the laboratory require the use of a graph for proper interpretation. Constructing a good graph can make a huge difference in the quality of a lab analysis, so it is worth the small amount of planning that good graphing involves.

As an example, look back at Figure IN-4. This figure shows how notebook results, instrument readings, and chemical calculations are used to construct a calibration curve. The principles that went into the graph are:

- **Proper identification of dependent and independent variables.** In most experiments, we can identify a variable that is changed deliberately, and we call this the independent variable. The values of the independent variable are plotted on the x-axis of the graph. Another variable will be measured to see how it changes in response to the independent variable. This is the *dependent* variable, and its values are plotted on the y-axis. In the case of Figure IN-4, a calibration curve describes how the amount of light absorbed by a sample (absorbance, abbreviated A) *depends* on the concentration of a dye (in this case, in units of moles per liter, or molarity). In the notebook records we see that the student actually recorded the volume of stock solution of dye used in each sample. She then *calculated* the concentration, which is the independent variable. The absorbance is the dependent variable.

- **Proper labeling of graph.** A graph presents a large amount of information, sometimes even an entire day's work, in one compact form. A *title* should indicate the nature of the graph (calibration curve), the system studied (carmine indigo), and, sometimes, the origin of the data (date and name). The *axis label* should spell out the variable and give the units. The axes should also indicate the values of the variables. These values may be integers, decimal, or scientific notation.

- **Scales.** A common error in graphing is the use of inappropriate scales for the axes. Two issues are important: the inclusion of an origin point and proper use of available space. In some experiments, a point where the values of both variables are zero should be included. This is the case with the calibration curve, where we expect that the absorbance will be zero when the concentration is zero. Take a moment to look at the available graph paper. What is the largest value you have for a variable? What is the smallest value (it may not always be zero). How does the range of values relate to the number of major grid points on the paper? Try to label and scale the axes so they fit the grid neatly while using as much of the paper as possible. The grid lines do *not* have to be integers, or even decimals. Scientific notation can be used, as is done with the *x*-axis (concentration) in Figure IN-4.

- **Data points.** The data points that relate the independent and the dependent variable form an *ordered pair* of data. Each ordered pair should be marked clearly on the graph by an obvious dot or square.

- **The curve or line.** The data points are the measured data for a graph, but sometimes, as with a calibration curve, we are more interested in what these indicate about other values that we have not measured. If that is the case, it is important to graph a good curve or line that shows how the data points relate to each other. It is generally poor practice to simply "connect the dots" in the graph. Instead, a smooth curve or line that fits all the data should be drawn. When the line is *between* measured points, then it should be a solid line. When the line extends *beyond* the measured data (including to the origin), then it should be a *dotted* line.

Interpreting Graphical Data

Examining a graph can reveal the relationship between the dependent and the independent variables, or allow a known relationship to be interpreted. But before the data are interpreted, the graph should be examined to determine if the data themselves are reasonable. In beginning science courses, you are likely to see data that have a smooth relationship between the variables. If one or more data points lie well away from the line or curve indicated by the others, then you may have an erroneous data point. That is why it is essential in many experiments to draw out a preliminary graph during the lab, when such data can be recollected.

In interpreting the graph, we seek to find a relationship that suggests something that physically or chemically connects the variables. There are three kinds of relationships you are likely to find between dependent and independent variables.

The value of the dependent variable may not change as the value of the independent variable changes. The graph appears "flat" as we move from left to right. There is *no dependence* in this case.

If the data points lie on or close to a line, then the value of the dependent variable has a *linear dependence* on the independent variable. The simplest case of a linear dependence is a simple proportionality: When we double the value of the independent variable, the value of the dependent variable also doubles. In that case, we expect that the origin (0, 0) will also lie on the line relating the variables. In other cases, the data may lie on a line with a formula $y = mx + b$, where m is the slope and b is the y-intercept of the line (where the line crosses the y-axis).

When linear dependence is present, we can use the data to get a line by comparing the values of two data points or two points on the line that we draw. In that case, we use the formula

$$m = \frac{y_2 - y_1}{x_2 - x_1}$$

This formula can be summarized as "rise over run," where "rise" is the change in the y-values and "run" is the change in the x-values.

The rise over run method suffers because it really uses only two points on the line. A better method is to include *all* the data points in a linear regression. This is a statistical procedure that minimizes the sum of the squares of the distance from the actual data to the line (which is why it is often called the "least squares" method). Regression analysis is available in many hand-held calculators and computer graphing programs.

If the data points do not lie on or close to a line but do smoothly change as the values of the variables change, then the value of the dependent variable has a *nonlinear dependence* on the independent variable. It is difficult to determine by eye what mathematical function applies in these cases. Although there are computer programs that can "fit" a curve with different functions, the most meaningful procedure is to try several different functions by regraphing the data with the y-values transformed. For example, in some experiments there is a logarithmic dependence between the dependent and independent variables. In that case, we would find that if we graph ln y on the y-axis, we get a line.

Calculating Unknowns from Graphs

Many experiments use graphs of known data points to get information about an unknown quantity. This is most important with calibration curves, which are created from data for known values of an independent variable and an observed value of a dependent variable. A good calibration curve can be used to determine what the "unknown" independent variable is in a sample. This is done by obtaining a reading for the dependent variable and then locating the value of the dependent variable that corresponds to that reading.

In using calibration curves, it is critical that you *interpolate* the data if at all possible. This means that the curve should only be used on values that are *within* the "known" data points. If, for example, we found in the experiment shown in Figure IN-4 that an unknown sample had an absorbance of 0.50, we would be able to determine that the concentration of dye in that sample had a value of 4.0×10^{-5} mol/L.

D. Analysis of Chemical Reaction Rates and Reaction Kinetics

The kinetics of a reaction are studied by following any one of the reactants or products through the course of the reaction. Consider the following case:

$$a\text{A} + b\text{B} \rightarrow c\text{C} + d\text{D}$$

The rate of this reaction can be studied by following the formation of the products or the consumption of the reactants. Rate is generally expressed as a change in concentration (usually molarity, mol L^{-1}) per unit time (usually seconds). To make valid comparisons among the different concentration changes, it is necessary to divide each concentration change by the stoichiometric amount of product:

$$\text{rate} = \frac{1}{c}\frac{\Delta[\text{C}]}{\Delta t} = \frac{1}{d}\frac{\Delta[\text{D}]}{\Delta t}$$

We can also express the rate as a function of the decrease in the concentration of a reactant. Because the concentration of a reactant decreases with time, changes in reactant concentration are negative. Therefore, we include a negative sign in the equation.

$$\text{rate} = -\frac{1}{a}\frac{\Delta[A]}{\Delta t} = -\frac{1}{b}\frac{\Delta[B]}{\Delta t}$$

Rate Laws

The rate of a chemical reaction depends on many factors, including the concentrations of chemical substances in the reaction mixture. This is generally expressed as a rate law, which relates the rate to the concentrations of substances:

$$\text{rate} = k[A]^x[W]^y$$

The rate law is experimental and is found by observing how the rate depends on concentration.

Here, A is the reactant in the decomposition reaction and W is the concentration of some other species. W can be another reactant. W can even be a substance that is not involved in the stoichiometry of the reaction. In that case, W is referred to as a *catalyst* or as an *inhibitor*.

As an example of the derivation of a rate law, consider the data shown in this table:

[A] (mol L^{-1})	0.648	0.555	0.441	0.277
Time (s)	10 s	30 s	60 s	120 s
Slope (mol L^{-1} s^{-1})	−0.0050	−0.0045	−0.0035	−0.0022
Rate (mol L^{-1} s^{-1})	0.0050	0.0045	0.0035	0.0022

The rate decreases as [A] decreases. When [A] drops from 0.648 M to 0.441 M (a change of −0.207, or −32%), then the rate decreases from 0.0050 to 0.0035 (a change of −0.0015, or −30%). The relative decrease in the concentration of A and the rate of the reaction are nearly equal, and we conclude that the rate law has an [A]^1, or simply an [A], term.

Thus, we can write, for just [A]: rate = k'[A]. This gives us the chance to determine the rate constant at some concentration of [A], k' = rate/[A]. Choosing the data point for 30 s, when [A] = 0.555, we get k' = 0.0045 mol L^{-1} s^{-1}/0.555 mol L^{-1} = 0.0081 s^{-1}.

This works fine for [A] and other substances in the reaction. But what about reactions where the rate depends on substances that are *not* involved in the reaction? Such substances, if they speed up a reaction, are called *catalysts*. Catalysts are not consumed or made in a reaction. Therefore, their concentration stays the same during the reaction.

We evaluate the order of a reaction for other components by examining the rate as we change the amount of the other components. Let's say we derived the dependence of rate on [A] with 0.0025 M catalyst W present. But what if we increase the amount of W to 0.0050 M? We then might get data that look like those in this second table:

[A] (mol L^{-1})	0.600	0.441	0.277	0.110
Time (s)	10 s	30 s	60 s	120 s
Slope (mol L^{-1} s^{-1})	−0.0093	−0.0074	−0.0046	−0.0022
Rate (mol L^{-1} s^{-1})	0.0093	0.0074	0.0046	0.0018

By comparing the data, we can make two important observations. First, the reaction with 0.0050 M W is much faster. Note that it now takes only 30 s to reach [A] = 0.441 M, whereas in the first case it took 60 s: doubling the amount of W doubles the rate. Second, note that the rate when [A] = 0.441 M increases to 0.0074 mol L^{-1} s^{-1} from 0.0035 mol L^{-1} s^{-1}. The rate of the reaction is twice that in the first case. This shows the reaction rate is dependent on [W]^1, or simply [W].

We now can write the rate law:

$$\text{rate} = k[A][W]$$

Information about the dependence on [W] can also be obtained by getting the rate constant k' for this second set of data. At 30 s, [A] = 0.441 mol L^{-1} and k' = 0.0074/0.441 = 0.0167, twice the value at [W] = 0.0025. This also indicates that the rate is linearly dependent on [W].

Integrated Rate Laws

Mathematical calculus provides a final and probably the most powerful method for determining the rate law. Let us return to the case of a rate law that depends on a single concentration, or on a single concentration that varies during a reaction. In that case, we can use calculus to determine the concentration of the species A as a function of time.

Reaction order	Rate law	[A] vs. t dependence	Characteristic plot
Zero	rate = k'	[A] = [A]$_o$ − kt	[A] vs. t Slope is −k
First	rate = k'[A]	[A] = [A]$_o e^{-kt}$ ln ([A]) = ln([A]$_o$) − kt	ln ([A]) vs. t Slope is −k
Second	rate = k'[A]^2	1/[A] = (1/[A]$_o$) + 2kt	1/[A] vs. t Slope is 2k

Figure IN-5 shows three plots for the data tabulated earlier.

Only one of the plots shows a straight line: the middle one. This indicates that the reaction is also first order in [A]. And a comparison of the two slopes confirms that the reaction is also first order in [W].

Figure IN-5

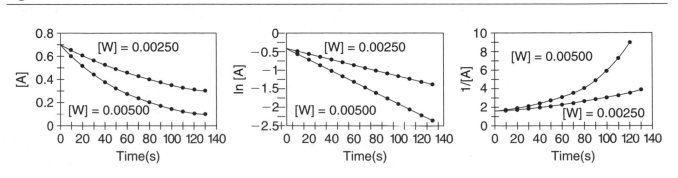

Designing a Kinetic Study

With these principles in mind, we can now outline four steps in designing a kinetic study:

1. Carry out the reaction in a manner that permits the periodic determination of a reactant or product concentration. Make sure that the experiment allows for fast and complete mixing of reactant solutions.
2. Collect the data until at least 25% or more of the reaction has proceeded. Repeat the data collection until it is reproducible.
3. Convert the experimental data into concentrations of a reactant. Calculate [A], ln[A], and 1/[A] and then graph each versus time to see which fits a line best. Determine k for the reaction from the best-fit line.
4. If a catalyst or other "unreactive" substance is present, vary it and see how k for the reaction changes with time.

Acknowledgments

The scenario for Experiment Group C was drafted by Prof. John Regalbuto in the Department of Chemical Engineering at the University of Illinois at Chicago.

The scenario for Experiment Group D and guidance for the adaptation of clinical methods to measure and treat unhealthy serum iron levels were provided by Prof. John Fitzloff of the College of Pharmacy at the University of Illinois at Chicago. Editorial assistance by Prof. Barbara Weil from William Rainey Harper College is gratefully acknowledged.

The scenario for Experiment Group E and guidance for the adaptation of the BCG method for determination of serum albumin were provided by Prof. Chuck Woodbury of the College of Pharmacy at the University of Illinois at Chicago. Editorial assistance by Dr. Audrey Hammerich is gratefully acknowledged.

The scenario for Experiment Group F and guidance for the adaptation of leaf compost chambers for CO_2 analysis from the decay of leaves were provided by Prof. John Lussenhop of the Department of Biological Sciences at the University of Illinois at Chicago. Editorial assistance by Dr. Helene Gabelnick from Harold Washington College is gratefully acknowledged.

The scenario for Experiment Group G and guidance for the description of pH methods in medicine were provided by Prof. Janean Holden of the College of Nursing at the University of Illinois at Chicago. Comments about the procedure by Dr. Dennis Lehman from Harold Washington College are gratefully acknowledged.

The scenario for Experiment Group H was drafted by Prof. John Regalbuto in the Department of Chemical Engineering at University of Illinois at Chicago. Editing support by Prof. Luke Hanley of University of Illinois at Chicago is gratefully acknowledged.

We gratefully acknowledge the assistance of Dr. David M. France in the development of the scenario of Experiment Group J and for providing background information on the factors engineers might consider in the design of a fire-resistant box. Dr. France is Associate Dean of Engineering and Professor of Mechanical Engineering at the University of Illinois at Chicago.

The scenario for Experiment Group K and guidance for the adaptation of procedure to analyze for the mineral nitrogen content in soil were provided by Prof. John Lussenhop of the Department of Biological Sciences at University of Illinois at Chicago.

The scenario for Experiment Group L was developed in conjunction with Dr. Robert Stahelin, a biochemist studying similar protein interactions with ions, small molecules, and other proteins.

The scenario for Experiment Group M was developed with Mr. Daniel Zavits, a physical chemist with extensive experience in alternative energy systems.

Experiment Group A

Methods of Inquiry and Measurement: Physical and Chemical Properties

Copper wire reacts with aqueous $AgNO_3$, giving the characteristic blue color of Cu^{2+} ions in solution while plating out silver atoms. (Peticolas/Megna/Fundamental Photographs.)

Purpose   These experiments will introduce you to the basics of mass and volume measurement in a chemical laboratory as you investigate some physical and chemical properties. They are all built upon some student inquiry, so they also introduce inquiry methods. By inquiry, we mean asking questions about how best to do an experiment and how to understand the results. Finally, the experiments will provide the opportunity to work individually and as a member of a group.

Schedule of the Labs

Experiment 1: Physical Properties: The Glass Bead Lab
- Determine the mass, volume, and density of both regular and irregular solids (individual work).
- Determine the mass, volume, and density of an individual glass bead (individual work).
- Combine beads with those of another group of students and determine the average volume, mass, and density (group work).

Experiment 2: Chemical Properties: The Activity Series
- Make detailed observations about chemical properties (group work).
- Determine an activity series for some metals (group work).

EXPERIMENT 1
Physical Properties: The Glass Bead Lab

Pre-Laboratory Assignment   **Due Before Lab Begins**

NAME: _____

Complete these exercises after reading the experiment but before coming to the
laboratory to do it.

1. Identify the following properties of calcium metal (listed in the reference book
   The Merck Index) as either chemical or physical properties and as intensive or
   extensive properties.

   (a) Lustrous silver-white surface (when freshly cut).

   (b) Ignites in air when finely divided, then burns with crimson flame.

   (c) Acquires bluish-gray tarnish on exposure to moist air.

   (d) Boiling point is 1440 °C.

   (e) Melting point is 850 °C.

   (f) Reacts with water, alcohols, and dilute acids with evolution of hydrogen.

2. Calculate the volume of a solid sphere if its circumference is 5.5 cm.

3. Determine the density of the solid sphere from Question 2 if it has a mass of
   7.421 g.

4. A strip of metal with dimensions of 5.0 cm $\times$ 2.0 cm $\times$ 0.2 cm has a mass of 38.64 g. Find the density of the metal and its identity from the following list that gives density values in units of g/cm^3: iron (7.86), copper (8.96), silver (10.5), gold (19.32), platinum (21.4).

5. How is water displacement used to find the volume of a solid? Name one important property the solid must have if this method is used.

6. You are not allowed to eat or drink in the lab at any time, including during this experiment, even though you are working with only glass beads and water. Explain why you are not allowed to eat and drink in the lab.

EXPERIMENT 1
Physical Properties: The Glass Bead Lab

Background

Chemical and physical properties are used to characterize and to help identify a substance. A chemical property describes how a substance interacts with other substances; it involves a possible change in the chemical composition when the property is observed. An example of a chemical property of sodium metal is its reaction with water to produce sodium hydroxide and hydrogen gas. A physical property describes the substance itself; no change in chemical composition occurs when the property is observed. Examples of physical properties of sodium metal include melting point, 97.82 °C; boiling point, 881.4 °C; and density, 0.968 g cm^{-3}.

Physical properties can be further categorized as extensive or intensive properties. Extensive properties depend upon the amount of matter present; intensive properties are independent of the amount of matter present. For example, the length of a piece of copper metal depends on the amount of matter present and so is an extensive property. However, the temperature at which water boils is 100 °C at standard atmospheric pressure regardless of the amount of water present; boiling point therefore is an intensive property.

Purpose

You will work individually and as part of a group during this experiment. Your goals are to characterize some properties of a substance and to develop expertise in (a) using common laboratory equipment and (b) making measurements of length, volume, and mass. You will practice some basic measurement techniques and error analysis, and you will be introduced to inquiry methods in chemistry: You will decide which procedures to use in certain cases.

Procedure

Part I: Formation of Groups

This experiment is best done in groups of two or three students. The group will decide how best to measure the volume of each of the solids used in the experiment, by the displacement method or by direct measurements. To measure volumes by displacement, partially fill a graduated cylinder with water and record this initial volume of water in your lab notebook. Add the solid, making sure it is completely submerged in the water. Record the final volume of the water with the solid submerged. The difference between this and your initial measurement is the volume of the solid.

The formulas for calculating the volume of some regular solids are as follows:

$$V_{sphere} = (4/3)(\pi r^3)$$
$$V_{cylinder} = \pi r^2 h$$
$$V_{rectangular\ solid} = l \times w \times h$$

where r is the radius, h is the height, w is the width, and l is the length.

Part II: Determining the Density of a Solid (Individual Work)

The density of a substance is the proportion that relates mass and volume:

$$d = m/V$$

To calculate the density of a substance, you need to measure its mass and volume. Obtain a block and a nut, bolt, or other solid as specified by your instructor. Record the measurements necessary to calculate the density of both solids. Be careful that all measurements are written with the precision dictated by the measuring device.

Part III: Determining the Density of a Single Glass Bead (Individual Work)

Obtain a single glass bead. Measure its mass on a balance. Determine its volume using the procedure decided upon by your group. Be sure to record this procedure in your lab notebook.

Using your values for mass (m) and volume (V), calculate the density (d) of the glass bead in grams per cubic centimeter.

Repeat your measurements two more times. Tabulate all of your data. Determine an average mass, volume, and density, including average deviations for your individual bead.

Part IV: Determining the Density of a Set of Glass Beads (Group Work)

Reconvene your group and add each of the glass beads to a set of additional glass beads received from your instructor. Obtain the mass of the entire set of glass beads.

To determine the volume of the set of glass beads, use the displacement method described in Part I. Determine the density of the glass beads. Repeat your measurements two more times. Report an average mass, volume, and density for your set of glass beads, including average deviations.

Report

1. Prepare a table that contains the density values for the block and the other objects obtained by all members of your group and those of one other group. Include the average density and standard deviation for each solid.

2. Describe the procedure you used to determine the volume of the individual glass bead. Tabulate the masses, volumes, and densities of the individual glass beads for each person in your group. Next, tabulate the masses, volumes, and densities of the *set* of glass beads. Explain which average density you think is a more reliable measure of the density of the glass beads—that of the individual beads or that of the sets of beads. Compare the masses, volumes, and densities of the individual glass beads with each other and with those of the *set* of glass beads. Identify each of the three properties (mass, volume, and density) as either intensive or extensive. Support your answers with your experimental data.

Questions to Answer in Your Report

1. Distinguish between the mass and the weight of a substance.

2. Why is the displacement method used to determine the volume of the set of beads instead of simply pouring the set of beads directly into the empty graduated cylinder?

EXPERIMENT 2
Chemical Properties: The Activity Series

Pre-Laboratory Assignment   **Due Before Lab Begins**

NAME: _____

Complete these exercises after reading the experiment but before coming to the laboratory to do it.

1. Which of the following properties are chemical properties?

   (a) Iron rusts in damp air.

   (b) Iron melts at 1535 °C.

   (c) Ethanol (C_2H_5OH) evaporates at room temperature.

   (d) C_2H_5OH burns to form CO_2 and H_2O.

2. Predict the possible products of these single displacement reactions.

   (a) Pb (s) + $Mg(NO_3)_2$ (aq) →

   (b) Ag (s) + $Zn(NO_3)_2$ (aq) →

   (c) Sn (s) + $Cu(C_2H_3O_2)_2$ (aq) →

3. Suppose that you want to test and record the reactivities of Mg, Ca, and Ba with HCl, NaOH, H_2O, and H_2SO_4. The following blank data table contains more columns and rows than you will probably need for this experiment. Label the columns and rows of the table in an efficient manner that makes sense to you for this experiment. (There is not one "right" way to do this.) Cross off any squares that represent duplicate combinations of reactants.

4. Look up the properties of H_2 (g), O_2 (g), and CO_2 (g) in the *Merck Index* or some other chemical handbook and use those properties to predict what will happen when a burning wooden splint is placed into separate test tubes—one filled with O_2, one filled with H_2, and one filled with CO_2.

5. Use the following experimental results to rank three metals (magnesium, aluminum, and zinc) from highest to lowest activity.

   (a) $Mg + Zn(NO_3)_2$ Reaction occurs

   (b) $Al + Zn(NO_3)_2$ Reaction occurs

   (c) $Al + Mg(NO_3)_2$ No reaction occurs

EXPERIMENT 2
Chemical Properties: The Activity Series

Background

The *chemical properties* of a substance describe how that substance reacts—or doesn't react—with other chemical substances. Some common chemical properties include the ability to react with water, oxygen, or other elements; flammability; decomposition into simpler substances; and the ability to act as an acid or base.

Chemical properties must be determined by testing a substance with various classes of chemicals to see whether any chemical reaction occurs. This is the tricky part. How do we know when a reaction has taken place? Many of the tests depend on our ability to "see" the results. For example, we recognize a chemical reaction by looking for the formation of a new substance.

We already know how to do this by looking at the words used in the description of the reaction or by examining a chemical equation to find the arrangements of atoms in the reactants and products. For example, we recognize that a chemical reaction has taken place in the equation

$$2 \, H_2O \, (l) \rightarrow 2 \, H_2 \, (g) + O_2 \, (g)$$

because the liquid H_2O has chemically changed into gaseous H_2 and O_2. But how can we tell whether a reaction has taken place when we combine unknown substances in the laboratory and do not have a chemical equation to examine? How do we know when one or more substances with chemical compositions different from the starting materials have been produced? As we will see, sometimes it is obvious that a reaction has occurred and other times it is not.

Chemical Properties: What to Look for in a Chemical Reaction

To assign chemical properties, then, we must look at some common chemical reaction types. The types of reactions in this experiment generally occur quickly and with noticeable product. Sometimes no reaction will occur. That's all right. *Failure* to react is also a chemical property and should be recorded as such. When you notice one of the signs that signal a chemical reaction, try to identify the products. For instance, if a gas is produced, can you figure out the name of the gas by looking at the reactants? What is reacting? What changes do you observe? Are these physical or chemical changes? What products are formed? Will similar compounds containing other elements from the same group on the periodic table react in the same way? These are important questions for a chemist to address.

Recognizing Signs of a Chemical Reaction

A. Change in Temperature

What makes chemical reactions occur in the first place? One force that drives chemical reactions is the tendency toward greater stability through minimization of energy. This means the reaction releases energy, often as heat. Such reactions are called *exothermic*. In other cases, heat is *absorbed* in what is called an *endothermic* reaction, so sometimes we experience a cooling sensation. Because this experiment uses very small amounts of reactants, it may be difficult for you to notice these temperature changes. However, you should look for—and, if observed, record—temperature change as a sign that a chemical reaction has occurred.

B. Formation of a Gas

Some compounds produce a gas during chemical reactions. For example, many compounds that contain SO_3^{2-}, CO_3^{2-}, or NH_4^+ ions often generate SO_2, CO_2, and NH_3 gases. Notice the similarity between the formulas of the anions and of the gases. Gas formation is often the result of heating, reaction with water, or reaction with an acid such as HCl. For example, calcium carbonate produces carbon dioxide gas when heated.

$$CaCO_3 \ (s) + heat \rightarrow CaO \ (s) + CO_2 \ (g)$$

In this case, we see from the reactant formula that the product gas must be CO_2. Indeed, CO_2 gas is often produced in reactions involving carbonates. In a similar manner, we would expect to get SO_2 gas when sodium sulfite (Na_2SO_3) decomposes by heating or, as shown below, when it reacts with water.

$$Na_2SO_3 \ (s) + H_2O \ (l) \rightarrow Na_2O \ (aq) + H_2SO_3 \ (aq) \rightarrow Na_2O \ (aq) + H_2O \ (l) + SO_2 \ (g)$$

Here we see an initial double displacement reaction (first step in the equation), followed by the decomposition of H_2SO_3 (sulfurous acid) into H_2O and SO_2. A similar—and more familiar—acid decomposition is that of carbonic acid, a component of carbonated beverages.

$$H_2CO_3 \ (aq) \rightarrow H_2O \ (l) + CO_2 \ (g)$$

In this case, we are aware of the escaping bubbles and the resultant "flatness" of the beverage after a time.

Another example of a chemical reaction that produces a gas is the reaction of some metals with acid. For example, when a piece of zinc metal is put into a solution of hydrochloric acid (HCl), hydrogen gas is produced.

$$2 \ HCl \ (aq) + Zn \ (s) \rightarrow ZnCl_2 \ (aq) + H_2 \ (g)$$

In this reaction, zinc displaces hydrogen from HCl, forming $ZnCl_2$ and H_2. You will find it easy to recognize the formation of a gas in today's experiment if you look for the bubbles of gas as they are produced in the aqueous solution.

C. Displacement of Metal Ions from Solution

If we put a strip of copper metal into a solution of $AgNO_3$, we will soon see bits of silver deposit on the copper metal.

$$Cu \ (s) + 2 \ AgNO_3 \ (aq) \rightarrow Cu(NO_3)_2 \ (aq) + 2 \ Ag \ (s)$$

This is a single displacement reaction in which copper displaces silver. What we "see" is that solid copper goes into solution as copper(II) ions while silver ions ($Ag^+ \ (aq)$) come out of solution as solid silver. Chemical equations should be scrutinized as to the physical state of each substance. This often provides a clue about the reactivity of a substance. These are reactions we can actually see. In this case, we should see the surface of the solid copper strip become pitted and irregular as it dissolves. The solution will begin to show the characteristic blue color of copper(II) ions in solution. We will also see some bits of silver forming.

If we test other metals for reactivity with the $AgNO_3$ solution, we would find that Mg and Ni give similar results but that there is no evidence of a reaction when metal strips of Pt or Au are used. Metals appear to have varying abilities to displace other metal ions from aqueous solutions of their compounds. This ability of a metal to displace other metals from aqueous solution is called its *activity*. More active metals displace less active metals.

Purpose

In this experiment, you will focus on making and recording detailed observations about chemical properties in reactions that give easily discernible results: change in temperature, evolution of a gas, or the appearance/disappearance of a metal in contact with a solution containing its metal ions. Because they are easy to see, these properties are often listed as *signs* of a chemical reaction. You will also test the reactivity of metals, keeping track of which metals react with particular solutions and which do not. Thus, you will have the information needed to rank the metals tested, from the most to the least reactive metal. This ranking is usually called the *activity series* for metals.

Procedure

Part I: Formation of Groups

You should work in pairs. For each part of this experiment, you should agree on how to set up the table needed to record your results. Each partner should be responsible for creating one of the tables needed. The reactions in this experiment require that very small amounts of chemicals be mixed in pairs using a 24-well plate. The tables should be set up to resemble the 24-well plate.

Part II: Preparation of Data Tables

The simplest way to organize the work you must do in this experiment is to create data tables in which to record your comments and observations about possible reactivities. Suppose you must test each of the solutions in Group A [$AgNO_3$, $Cu(NO_3)_2$, $Zn(NO_3)_2$, $Mg(NO_3)_2$, and $Fe(NO_3)_3$] with each of the solutions in Group B ($NaCl$, Na_2S, $NaOH$, and Na_2CO_3). To create a data table, write the formulas of the Group A chemicals across the top row and the formulas of the Group B chemicals down the first column of the table. (Note that this assignment is arbitrary; the positioning of the groups of chemicals could be reversed.) The table should look something like this:

TABLE 1

Observations of Results of Mixing Group A Chemicals with Group B Chemicals

	$AgNO_3$	$Cu(NO_3)_2$	$Zn(NO_3)_2$	$Mg(NO_3)_2$	$Fe(NO_3)_3$
NaCl					
Na_2S					
NaOH					
Na_2CO_3					

The setup of the table resembles the 24-well plate in which you will mix solutions. Mix small amounts of the two solutions that intersect at each position on the data table—that is, mix small amounts of NaCl and AgNO$_3$ in the first well, etc. Carefully observe each combination of solutions for signs of a reaction, then record your observations in the data table. The key here is to carefully assemble a data table so that you will know which combinations of chemicals react and which do not. Be careful not to let excess solution spill over into neighboring wells, as this may confuse your results.

Part III: Reaction of Metals with Acid (Group Work)

One partner should prepare a table correctly labeled for this part of the experiment while the second partner gathers the needed materials. You will need about 10 mL of HCl and strips of Zn, Cu, Fe, Mg, and Ag metals. When you both are ready to begin, fill 5 wells with HCl. Insert a different metal strip into each well of HCl solution and observe each metal for signs of a chemical reaction.

If you see evidence of a gas being produced, repeat the reaction in a test tube and, in a hood, test the gas with a burning splint. Record all of your findings in the table.

Part IV: Reaction of Carbonate Compounds (Group Work)

Obtain about 10.0 mL of HCl and a small amount (about one spatula-full) of sodium carbonate, potassium carbonate, calcium carbonate, and magnesium carbonate. This time, the second partner should prepare a table such that each carbonate salt is separately tested for reaction with both water and HCl. Your observations should include the solubility of the substance in water and in HCl, the formation of gas bubbles, the relative rate of bubble formation, and the evolution of heat.

When both partners are ready, add small amounts of each substance to the portions of water that have been put into separate wells of the 24-well plate; observe for signs of a chemical reaction. Next, add small amounts of each substance to the separate portions of HCl solution and observe for signs of a chemical reaction. If you see evidence of a gas being produced, repeat the reaction in a test tube and, in a hood, test the gas with a burning splint. Record all of your findings in the table.

Part V: Activity of Metals (Group Work)

Decide the tasks of table preparation and material gathering. Prepare a table for your observations. Obtain strips of the following metals: Ag, Cu, Fe, Mg, and Zn. Also obtain about 10 mL of each of the following solutions: AgNO$_3$, Cu(NO$_3$)$_2$, Zn(NO$_3$)$_2$, Mg(NO$_3$)$_2$, KNO$_3$, and Fe(NO$_3$)$_3$.

Test each of the five metals in each of the solutions and observe for signs of a chemical reaction. Record all of your findings in the table.

Report

Write a summary of the experiment, including a description and the data tables for each part of the experiment. What chemical properties did you study? Did all of the combinations of reactants result in a chemical reaction? Can you suggest a reason why some things reacted while others did not? Give examples to support your suggestions.

Questions to Answer in Your Report

1. Which of the metals in Part III reacted with the acid? Which did not? Discuss the identity of the gas evolved and give evidence to support your identification.

2. Record the relative reactivity of the carbonate salts (Part IV) in water and in HCl (use the rapid or slow formation of bubbles as your indicator of reactivity). Discuss the identity of the gas formed and give evidence to support your identification. Can you suggest a reason for any differences in reactivity that you observed?

3. Which pairs of reactants underwent a chemical reaction in Part V? Which showed no signs of reaction?

4. Examine the reactions that occurred in Part V and formulate a list of the metals from the most reactive to the least reactive. Look at your observations for Part III and predict where H_2 would rank in this activity series of metals.

Experiment Group B

Measurement of Chemical Ratios: Float an Egg

The density of a solution affects whether an object floats in that solution. (Jeff. J. Daly/Visuals Unlimited.)

Purpose These experiments will introduce how chemists use proportions in measurement. They allow you to investigate how measuring one thing allows you to determine the properties of other things indirectly.

Schedule of the Labs

Experiment 1: Use of Volumetric Glassware: The Floating Egg Problem
- Prepare a solution to float an egg (group work).
- Determine solution density using volumetric glassware (individual work).

Experiment 2: Chemical Proportionality: Carbonate and Hydrochloric Acid
- Characterize the reaction of a carbonate with acids (group work).
- Determine the mass-to-volume ratio for the reaction of a carbonate with hydrochloric acid (group work).
- Determine the concentration of an unknown HCl solution (individual work).

EXPERIMENT 1
Use of Volumetric Glassware: The Floating Egg Problem

Pre-Laboratory Assignment   **Due Before Lab Begins**

NAME: _____

Complete these exercises after reading the experiment but before coming to the laboratory to do it.

1. List four things to avoid doing when using the balance.

2. You are in a hurry to dry some glassware. Comment on the advisability of the following methods.

    (a) Use paper towels.

    (b) Use compressed air from the lab air line.

    (c) Use acetone of uncertain cleanliness.

    (d) Heat in an oven.

    (e) Use clean acetone and then heat in an oven.

3. What effect will there be on the accuracy of a measurement if you use a volumetric pipet that is not completely dry but instead contains a few drops of the same liquid that you will be drawing into the pipet?

4. Define meniscus. State how you use the meniscus in making volumetric measurements.

5. Label the types of glassware shown in Figure B-1.

Figure B-1

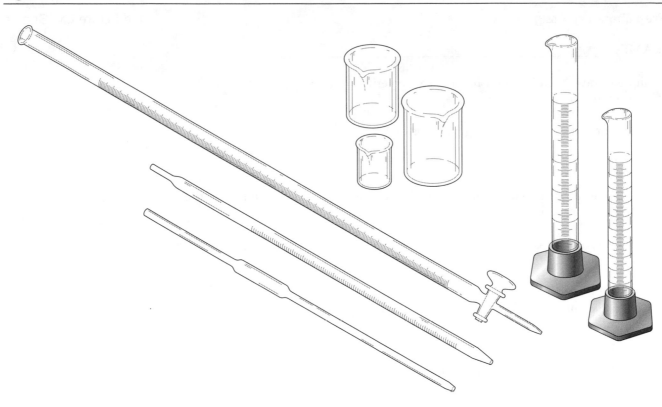

6. Although you are working with NaCl solutions instead of NaOH solutions in this experiment, it is not safe to pipet by mouth (this is never an acceptable technique). Explain.

EXPERIMENT 1
Use of Volumetric Glassware: The Floating Egg Problem*

Background

We are all familiar with direct measurements of things. If, for example, we weigh a 1.0-cm³ cube of iron and find it has a mass of 7.86 g, we have determined directly from a piece of iron that the density of iron is 7.86 g cm⁻³. But in many cases we cannot measure a property directly. Therefore, we resort to indirect measuring. We make a direct measurement and, using a known relationship, apply it in determining another quantity we cannot measure directly.

In this lab, you will see how we can indirectly determine the density of an object by directly measuring the density of a solution in which it just floats.

To carry out good measurements, chemists use special *volumetric* glassware in the chemistry laboratory rather than ordinary beakers and flasks. Understanding the proper use and limitations of such glassware is important in achieving reliable results. Experimental technique and the inherent accuracy of the glassware both affect experimental results.

In this experiment, you will practice with volumetric glassware to improve your technique and to solve an important historical problem in practical chemistry. Volumetric glassware is designed for two quite different purposes:

- To contain (TC) an accurate volume of liquid
- To deliver (TD) an accurate volume of liquid

Although a container meant to *contain* an accurately known volume could be used to deliver that volume, it is generally not used to deliver because the amount of liquid actually delivered would never exactly equal the amount originally in the container. Some liquid would inevitably remain clinging to the interior walls of the container.

When we must know accurately how much liquid is delivered, we use TD glassware. Some liquid also clings to the walls of TD glassware, but the calibration of the TD glassware takes this into account.

A Historical Question

For many years, soap was made at home from a variety of recipes. Animal fat, usually from cattle (also called tallow), was cooked with a lye solution. Lye, though it was mostly sodium hydroxide, could not be made from purified chemicals, as it is now. Instead, the solution was obtained from ashes and water. The ashes were treated with hot water, and then the mixture was filtered to obtain a solution. But before this could be used in soap making, the concentration of the lye solution had to be checked.

One way to check the concentration was to determine the density of the solution. One simple test was to try to just float a raw egg in the solution. If the egg sank, the density of the lye solution was too low. If the egg floated too high, the density was too great, and water was added before adding the fat. To "just float" in this case means to make the top of the egg just touch the top of the solution, without any significant amount of the egg protruding above the surface of the solution. At that point, the

---

*Portions of this laboratory, including the historical question, are adapted from materials developed by Dr. Susan Hershberger in the Miami University Middletown in their Scenario Lab program.

density of the lye—and therefore its concentration—was correct. Thus, we can see how soapmakers had mastered the art of indirect measurement!

We will use this strategy in a different manner. We will use a solution that just floats an egg to determine the density of the egg.

Purpose

You are to determine the density (in grams per milliliter) of the solution needed to just float an egg. In addition, you are to determine in what way each of the following affects the value and accuracy of the calculated density: (a) the type of volumetric glassware used (volumetric flask, Mohr pipet, volumetric pipet, or buret) and (b) the degree of freshness of the raw egg.

Procedure

Part I: Formation of Groups and Preparation of the Solution

This experiment is best done in groups of three or four individuals. Follow the detailed directions of your laboratory instructor.

Because lye is caustic and corrosive, the substitute sodium chloride will be used in this lab. The solutions are still concentrated enough to pose a hazard to the eyes. Goggles must be worn. Salt, eggs, standard glassware, balances, and deionized water will be available. Other items may be available on request. The eggs are uncooked. But because they are used in a lab, and because they are at room temperature for long periods, they should *not* be reused as food.

Clean and dry all glassware before use.

Follow these guidelines to prepare a solution that will just float an egg:

- Make at least 1500 mL of the solution
- Use only one egg for testing. Note the code. At the end of the experiment, you will use this to find out the freshness of the egg from your instructor.

When you have prepared the solution, each individual should record the procedure in his or her notebook.

Part II: Measurement of Solution Density (Individual Work)

Each individual should use all four kinds of measuring devices.

The group should obtain a 25-mL buret, a 25-mL Mohr pipet, a 25-mL volumetric pipet, and a 25-mL volumetric flask. (*These must be cleaned and returned to their designated location at the end of the experiment.*)

Decide who will work first with each type of equipment. For the equipment that dispenses solution (pipets and burets), you will determine the mass of the solution by collecting the solutions into a weighed Erlenmeyer flask. For the volumetric flask you will determine the mass of the solution in the flask itself. So, obtain the weight of a clean, dry Erlenmeyer flask and of the volumetric flask. Don't forget to take a notebook to the balance, too. Before weighing, be sure the balance reads zero when nothing is on the pan. Check with your instructor if this is not the case. Then place the flask on the balance. This is the mass of the empty container.

Record the sensitivity of the balance in your notebook. Note for future reference that when the net mass of something is determined by subtracting the empty (tare) mass from the gross mass, *the uncertainty will be twice the sensitivity*. This is because there is no guarantee that each of the two mass readings will deviate from the true mass in the same manner (that is, both larger or both smaller).

Now, carry out three repetitions of a determination of the density of the solution with each piece of glassware by measuring a volume close to 25 mL of the solution and then determining the weight of the flask plus the solution.

A. Buret

Make sure the stopcock on the buret is properly assembled. Clean the buret and make sure that it drains smoothly with no beading of water on the walls. Then pre-rinse it with a little of your solution. Use a buret clamp and ring stand to mount the buret vertically. Fill it to about the zero level with your solution. Drain some out to ensure that no pure water or air gaps are in the tip of the buret. Record the level reading after it has stabilized.* The level should be recorded to the nearest 0.01 mL, but it does not have to be at exactly 0.00 mL. (Estimate the location of the bottom of the meniscus between the calibration marks, which are 0.1 mL apart.)

Deliver solution from the buret into the clean, dry, weighed Erlenmeyer flask. Deliver until the buret level is near the 24-mL mark. Record the reading after waiting for the level to stabilize.

Now weigh the flask plus the solution and record the total mass. When you subtract the mass of the empty, dry flask, you will obtain the mass of the solution.

Repeat the entire procedure, including rinsing the buret, for a total of three trials. You want to develop your technique so that you can use a buret accurately and reproducibly (precisely).

B. Mohr Pipet

Never apply suction by mouth to fill up this or *any* pipet. Instead, use pipet bulbs that will be available in the laboratory.

Using the 25-mL Mohr pipet, carry out the weighing procedure of Part A. Be sure that the pipet is clean and drains correctly, then pre-rinse it with some of your solution. Attempt to deliver exactly 23.00 mL of solution into the dry weighed Erlenmeyer flask. If you undershoot or overshoot, record whatever actual volume you finally deliver. Do a total of three trials. Attempt to perfect your technique.

C. Volumetric Pipet

Carry out the weighing procedure of Part A using a clean, pre-rinsed 25-mL volumetric pipet to deliver 25.00 mL of solution. If the pipet is marked TD, it is designed to give 25.00 mL by draining with gravity: Do not include the solution that remains in the tip after draining. The tips of the pipets are very fragile. Be careful not to exert pressure on them. Do a total of three trials.

D. Volumetric Flask

Fill your pre-weighed volumetric flask with the solution, stopping at the calibration mark. This will give you a volume of 25.00 ± 0.02 mL. Weigh the solution-filled container and record the mass. This is the mass of the flask plus the solution "contained," and, by subtracting the mass of the empty flask, you can determine the mass of the solution contained. *Empty* the contents of the container into the clean, dry, weighed Erlenmeyer flask. Record the mass of the Erlenmeyer flask plus the solution. From this you can calculate the mass of solution "delivered." Repeat this procedure three times.

Before you leave the lab, share your results with the others in your group. Also, obtain data from other groups with eggs of different freshness.

Cleanup.   The glassware should all be rinsed well with deionized water before returning it to its designated location; unrinsed glassware will dry out to leave solid

---

*Readings that are obtained too quickly do not include the liquid that slowly trails down the sides of the buret.

sodium chloride that may clog the tips. Also, return the eggs to the proper container for that code. Do not put all the eggs in one basket.

Report

Your report should include a full procedure and analysis of your data. This analysis should include an indication of the variation among the measurements. Also, present a comparison of your data with those obtained by the others in your group and by others in the room with the same glassware. How did the amount of variation in your data compare to the amount obtained by others with the same or with different glassware? Finally, answer these questions in one or two sentences:

1. What was the density of the solution in which the egg just floated?
2. How valid is the assumption that aqueous sodium chloride is an adequate replacement for aqueous sodium hydroxide in this problem? Explain your reasoning.
3. Compare the density you obtained with the densities obtained for different eggs. What, specifically, was the effect of the freshness of the egg?
4. What were the advantages and disadvantages of each measuring device?
5. Compare the mass of the solution contained in the volumetric flask to the mass of the solution delivered by the volumetric flask. Account for differences. Which of the two masses, contained or delivered, should you have used in the determination of the density of the solution? Why?

EXPERIMENT 2
Chemical Proportionality: Carbonate and Hydrochloric Acid

NAME: _____

Complete these exercises after reading the experiment but before coming to the laboratory to do it.

1. Suppose a group of students is to study the reaction between solution A and four different chemical reagents. Two methods can and should be used to characterize this set of reactions: Method 1 adds the chemical reagent by drops; method 2 weighs the amount of chemical reagent added.

   (a) What is the total number of tasks that must be completed by the group?

   (b) Devise a reasonable work schedule that divides the tasks equitably between two group members.

   (c) Devise a reasonable work schedule that divides the tasks equitably among four group members.

2. Suppose you read the volume of a solution contained in a graduated cylinder from three different eye levels, marked A, B, and C in Figure B-2. Label each eye-level position with the type of result you would expect: an accurate measurement, a high measurement, or a low measurement.

Figure B-2

3. Suppose you experimentally determined that a 22.6-mL sample of a solution required 3.08 g of KOH to completely react. What mass of KOH would be required to completely react with 83.3 mL of the same solution?

4. What is the slope of the line in Figure B-3? Include the units of the slope.

Figure B-3

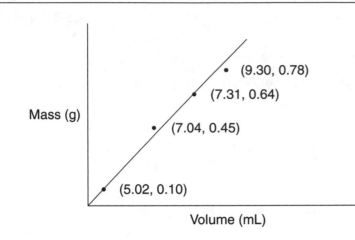

5. Suppose, as you are adding sodium carbonate to one of the acid solutions, some of the acid solution splashes out of the beaker and on to your hand. What will you do?

EXPERIMENT 2
Chemical Proportionality: Carbonate and Hydrochloric Acid*

Background

Some of the basic principles of chemistry were recognized and used well before the formulation of atomic theory in the early nineteenth century. Many of these practical procedures are still employed today. For example, some food preparation requires careful control of the amount of baking soda (sodium hydrogen carbonate) or baking powder (a mixture of tartaric acid and baking soda). Other examples involve the mixture of ingredients needed to make a good cement, or the proper application of fertilizers to adjust soil conditions. In all of these, there is an implicit rule: We can determine how much to add by considering the relative proportions of the components. This lab shows how an early method for determining acid concentration used such proportional reasoning.

A Historical Question

In the eighteenth century, leather tanners developed a simple method to determine the concentration of the acid they used in the tanning process. They added small portions of a solid carbonate salt to the acid solution, which resulted in fizzing as the carbon dioxide gas generated by the reaction escaped from solution. They continued adding the carbonate salt until there was no more evidence of fizzing. If they used a set volume of acid, they could compare the amount of salt required for various samples of acid and thus measure them against each other. Acid samples that required large masses of carbonate salt were more concentrated than acid samples that required small masses of carbonate salt.

Purpose

In this lab, you will investigate what occurs when a solid substance, a metal carbonate, is added slowly to a solution of an acid. You should be able to validate the proportionality of the amounts—volume and mass—that are required to accomplish this. This proportionality underlies the idea of chemical stoichiometry.

By completing this experiment you will:

- Gain experience making observations.
- Gain experience in designing procedures for a chemistry laboratory.
- Recognize some easily observable changes that signal a chemical reaction.
- Investigate proportional reasoning applied to mass and volume measurements.

Solutions are defined as homogeneous mixtures of solutes and solvents. In this course, you will be using many solutions, most of them water-based or aqueous solutions, in which water is the solvent. The amount of solute dissolved in a specified volume of water is given by the concentration: Dilute solutions contain small amounts of dissolved solute and more concentrated solutions contain greater amounts of dissolved solute.

---

*Portions of this laboratory, including the historical question, are adapted from materials developed by Purdue University.

How do we measure the concentration of a solution? We can use many different methods, some of which are more accurate than others. We can use qualitative or quantitative information that is based primarily on the physical properties of the solution.

We can also use chemical methods to determine concentration. These methods involve a chemical reaction between some reagent and the dissolved solute. Three things are required here: (1) We must carefully measure the amount of reagent added, (2) we must have some means of knowing when to stop adding the reagent, and (3) we must be able to relate the amount of reagent added to the chemical reaction being monitored.

In the experiment you will do today, a metal carbonate is added to the solution until there is no more evidence of carbon dioxide being generated. The chemical equation for this reaction is

$$M_2CO_3\ (s) + 2\ HCl\ (aq) \rightarrow 2\ MCl\ (aq) + CO_2\ (g) + H_2O\ (l)$$

Originally, the evolution of CO_2 gas was used to determine when the acid was used up. When addition of carbonate no longer produced CO_2 gas, the reaction was complete. To make your work more precise, we will also add an acid–base indicator, a substance that changes color dramatically when the reaction is complete.

To monitor the amount of metal carbonate required by the reaction, you can simply subtract the mass of carbonate you have at the end from the mass you had at the beginning. The relationship between the chemical amount (usually the number of moles) of metal carbonate and that of the HCl is given by the stoichiometry of the equation. Because we are working with mass measurements and unknown solution concentrations, we must *experimentally* determine the mass-to-volume ratio.

You have two kinds of measurements to record: mass and volume. Mass is easily obtained by using a balance. For this experiment, you will use the analytical balance, which will allow more precise measurements than centigram balances. Review the proper techniques for using the analytical balance. Do not bring any open containers of chemicals into the analytical balance room. Instead, weigh a clean, dry beaker, then return to the laboratory. Use the centigram balances to measure approximately 1 g of metal carbonate into the beaker, then return to the balance room and weigh again. The difference between the two masses is the mass of your initial supply of metal carbonate.

At the end of each experiment, you will weigh the beaker with the remaining metal carbonate. This is called "weighing by difference" and will allow you to determine the mass of metal carbonate actually consumed in the experiment. Report all masses to 0.001 g.

From your experience in the skill-building lab, choose an appropriate piece of glassware to measure solution volumes.

Procedure

Part I: Formation of Groups

This experiment is best done in groups of three or four individuals. Follow the detailed directions of your laboratory instructor. As a group, obtain a small amount of sodium carbonate to be used in Part II. The goal of Part II is to distinguish whether or not a gas is evolved when metal carbonate is added to water and to various water solutions. After completing Part II, develop, as a group, a work plan to complete the remainder of the experiment.

Part II: Dissolution or Chemical Reaction?
What Do Your Observations Mean? (Group Work)

1. Using a spatula, add a small amount of sodium carbonate to about 10 mL of water. Swirl the solution so that the solid has an optimum chance to dissolve. What observations can you make? Does the solid in fact dissolve? Do you sense any other changes? Add one drop of indicator and record the color.

2. Add a small amount of sodium carbonate to about 10 mL of the hydrochloric acid solution and swirl gently. Does the solid dissolve? Do you sense any other changes? Add one drop of indicator and record the color.

Part III: Measuring the Strength of a Solution (Individual Work)

This part of the procedure should be divided up among the individual group members according to the work plan decided upon by the group.

The group must determine the mass of sodium carbonate required to react completely with a measured volume of acid. You have three acid solutions of different concentrations to examine. The total volume of each solution that is available is limited to about 100 mL. Each group will perform four trials with different volumes for each acid solution, for a total of twelve trials.

Add one drop of indicator before adding any sodium carbonate. Compare the observed endpoint of the reaction as signaled by the end of CO_2 bubbling and the color change of the indicator.

Part IV: Collection and Analysis of Group Data (Group Work)

The next step in reporting results for any experiment is to examine both the procedure and the experimental results. Reconvene with the whole group to be sure you have each collected the necessary data. Carry out a preliminary analysis together. Failure to do this may mean that you will not have the required information to prepare your report.

Prepare tables of data for all three solutions. Each table should contain the four solution volumes your group used and the mass of sodium carbonate required for each volume. You will use these tables to analyze your data.

Do the following for each solution.

1. Compute the ratio of mass to volume for each of the four trials. Use units of grams and milliliters. Determine the average value of this ratio.

2. Prepare a graph of mass versus volume with the mass of the carbonate along the y-axis and the volume of the HCl along the x-axis. Determine the slope of the best line drawn through the data points. (If you use a computer to prepare this graph, you may use a linear regression analysis to get the slope; if you do, report the correlation coefficient the program reports.)

The three graphs that you prepare should show that mass and volume are proportional. This means that the data for each solution fit a straight line that has an origin at (0, 0). The slope of the line gives a ratio that is a constant of proportionality for each solution. The slope has the ratio of mass of sodium carbonate divided by volume of HCl. Note that this is not a density value. It is the ratio of one reactant (in grams) to a second reactant (in milliliters).

You will notice that the three lines, for the three different HCl solutions, are different from one another because the concentrations of the three HCl solutions are different. We can fix this by graphing the mass of sodium carbonate (on the y-axis) versus the moles of HCl (on the x-axis). Ask your instructor for the concentrations of the three acid solutions. To determine the number of moles of HCl in each sample, we multiply the concentration of HCl (in moles per liter) by the volume (in liters). Prepare a graph that compares mass of sodium carbonate with moles of HCl. Include data from all three acid solutions on the same graph, but use a different symbol for the data points for each acid solution.

Part V: Analysis of an Unknown Concentration of HCl (Individual Work)

As a group, you have prepared a graph of the mass of sodium carbonate, Na_2CO_3, versus moles of hydrochloric acid. You can now use this to determine the concentration of an unknown solution of HCl.

Ask your instructor for directions for obtaining an unknown. Record the unknown code or number. You will only get one portion of this, so use it carefully.

Measure out 25.0 mL of the unknown. Put it into a clean beaker and add a drop of indicator. Weigh a portion of sodium carbonate and begin adding it to the hydrochloric acid. Do this very carefully, to determine exactly how much sodium carbonate is needed for the reaction. Once you are certain the hydrochloric acid is completely reacted, reweigh your sodium carbonate.

You can use your data and the data collected by the group to determine the concentration of your unknown. Your group data will tell you the ratio of grams of sodium carbonate to moles of HCl for the known solutions. Because you now know how much sodium carbonate was needed for your unknown, you can use your graph to find the moles of HCl in your unknown. Divide this by the volume of unknown (0.0250 L) to determine the concentration of HCl.

Report

Summarize the procedure and your group's work plan for this experiment. Comment on any difficulties with the procedure as a means of accomplishing the goal of the experiment.

Your report must also include the data tables for the known solutions, the average mass-to-volume ratios, the slopes of the lines from the graphs of mass versus volume and mass versus moles (and the correlation coefficient if you used a statistical program to analyze the data), and the concentration of the unknown HCl solution. Include sample calculations for any calculated values.

Questions to Answer in Your Report

1. For each acid solution, how closely do the individually computed mass-to-volume ratios agree with one another?

2. For each acid solution, compare the slope of the line with the average value obtained for the computed mass-to-volume ratios for that solution.

3. Compare the average of the computed mass-to-volume ratios of the solutions. Rank the three acid solutions from most concentrated to least concentrated based on the mass-to-volume ratios. What is the relative ratio of the concentrations of the solutions? Compare your experimental results with the relative ratios of the concentrations calculated using the concentrations of the acids obtained from your instructor.

4. Is the mass of sodium carbonate proportional to the number of moles of HCl? Support your answer with experimental evidence.

5. Does the number of moles of HCl that react with a given mass of sodium carbonate depend on the concentration of the HCl? Explain.

Experiment Group **C**

Mass Stoichiometry

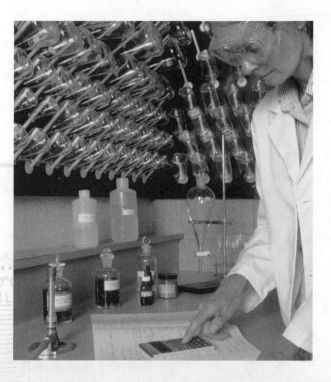

Calculating proper amounts of reactants is an essential part of many chemical syntheses. (Stephen Frisch/Stock, Boston/PNI.)

Purpose Chemical reactions, represented symbolically by chemical equations, result in the production of new substances. It is often important to understand reactions both qualitatively—what is involved—and quantitatively—how much. The measurement of the amount of substances involved in a chemical reaction is known as chemical stoichiometry. The coefficients in the balanced equation are the exact number of moles of reactants needed and the exact number of moles of product formed. The phrase *stoichiometric amounts* describes the case when the relative amounts of the reactants match the amounts required by the balanced equation.

When stoichiometric amounts of reactants are not involved, one of the reactants will run out first. This reactant is called the limiting reactant, or limiting reagent. The other reactant or reactants are said to be in excess. In this experiment group, you will explore reaction stoichiometry and the limiting reactant concept in the synthesis of a metal complex in the skill-building lab. Then you will apply those skills in the analysis of the amount of phosphorus in a fertilizer sample.

Schedule of the Labs **Experiment 1: Skill-Building Lab: Synthesis of Metal Complexes**
- Synthesize the metal saccharinate complex (individual work).
- Practice separation by filtration technique (individual work).
- Isolate the saccharinate complex (individual work).
- Compare the amounts of product formed by all members of the group to characterize the limiting reactant (group work).

Experiment 2: Application Lab: Gravimetric Analysis of Phosphorus

- Prepare the fertilizer sample for analysis (group work).
- Determine the percentage by mass of phosphorus in the fertilizer sample (individual work).

Scenario Farmers in the United States apply about 54 million tons of fertilizers annually. In addition, many consumers also use fertilizers in their home gardens. Whether for large- or small-scale use, fertilizers serve as sources of the primary plant nutrients (macronutrients) nitrogen, phosphorus, and potassium. Most consumer fertilizers contain all three macronutrients. For example, a fertilizer labeled 5-10-5 contains 5% nitrogen, 10% phosphorus, and 5% potassium by mass. However, this labeling system does not reflect the actual composition of the fertilizer. While the nitrogen is expressed as percent by mass of nitrogen, the phosphorus and potassium are expressed as percent by mass of P_2O_5 and K_2O.

Although these compounds are not present in the fertilizer, this labeling system for phosphorus and potassium is based upon the method used to assay fertilizers developed by Samuel Johnson, who is known as the "father of the fertilizer industry." In this method, phosphorus and potassium content were determined by heating the sample in air to high temperatures, a process known as *ashing*, resulting in the formation of phosphorus and potassium oxides.

For this application, you will assume the role of scientists at an analytical chemistry company. A leading fertilizer company has hired your laboratory to assay their fertilizer. The company has recently been a target of a consumer advocate group that questioned the composition of their fertilizers. Therefore, they agreed to have their product tested by an independent laboratory. Your group is responsible for determining the phosphorus content of the fertilizer. You decide to perform a gravimetric analysis to accomplish your task. Several of your colleagues will perform the analysis as well. It is standard practice in analytical laboratories for several experimenters to analyze different samples of the same substance to compare results and validate the procedure. Therefore, after you complete your analysis, you will compile the data from your entire group and prepare a report for the fertilizer company explaining your method and your results.

EXPERIMENT 1
Skill-Building Lab: Synthesis of Metal Complexes

Pre-Laboratory Assignment   **Due Before Lab Begins**

NAME: _____

Complete these exercises after reading the experiment but before coming to the laboratory to do it.

1. In this experiment, you will carry out a displacement reaction. A similar reaction occurs in the synthesis of complexes of metals with the sodium salt of dimethylglyoxime (NaDMG):

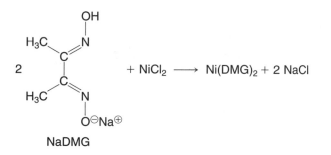

NaDMG

   (a) If you carry out this reaction with 2.00 g of $NiCl_2$, what mass of NaDMG will you need for the reaction?

   (b) What is the theoretical yield of $Ni(DMG)_2$ you expect from the experiment in part (a)?

   (c) If the actual reaction gives only 1.22 g of $Ni(DMG)_2$, what is the percentage yield of the reaction?

2. When copper(II) sulfate is prepared from copper(II) nitrate and sodium sulfate, the reaction initially produces a hydrated product:

$$Cu(NO_3)_2 + Na_2SO_4 + 5\ H_2O \rightarrow CuSO_4 \cdot 5\ H_2O + 2\ NaNO_3$$

If $CuSO_4 \cdot 5\ H_2O$ is heated, the water is removed, leaving anhydrous $CuSO_4$. What mass of anhydrous $CuSO_4$ will be produced from 2.20 g of $CuSO_4 \cdot 5\ H_2O$?

3. Indicate the safety concern with saccharin. Is this also a problem with sodium saccharinate?

EXPERIMENT 1
Skill-Building Lab: Synthesis of Metal Complexes

Background

The displacement reaction is among the most fundamental transformations in chemistry. In a displacement reaction, we substitute one atom or group of atoms for another. In its simplest scheme, this occurs with a *single* displacement reaction:

$$AX_n + nY \rightarrow AY_n + nX$$

A *double* displacement reaction can be viewed as an exchange of groups. This often happens with ionic compounds.

$$AX_n + nDY \rightarrow AY_n + nDX$$

Examples of double displacement reactions include

$$NaBr\ (aq) + AgNO_3\ (aq) \rightarrow AgBr\ (s) + NaNO_3\ (aq)$$
$$Na_2SO_4\ (aq) + BaCl_2\ (aq) \rightarrow BaSO_4\ (s) + 2\ NaCl\ (aq)$$

In the case of these reactions, we have a single product that is sparingly soluble in water. Therefore, the reaction is relatively simple: Mixing the solutions causes the immediate formation of a precipitate, and then that product can be isolated by filtering away the solution.

Other reactions, especially in organic chemistry and with most of the other transition metals, may occur in a similar fashion. But the rate of the reaction may be very slow. In those cases, there is plenty of time for other reactions to occur, giving undesirable side-products. We may need to heat the reaction mixture to complete the reaction in a reasonable time.

For example, the formation of soap requires a simple displacement by hydroxide of three long-chain molecules, called fatty acids, from a molecule of glycerol:

$$
\begin{array}{l}
CH_2\!-\!O_2CC_{17}H_{33} \\
CH\!-\!O_2CC_{17}H_{33} \quad + 3\ NaOH \longrightarrow C_3H_8O_3 + 3\ C_{17}H_{33}CO_2Na \\
CH_2\!-\!O_2CC_{17}H_{33}
\end{array}
$$

This reaction can be done by cooking animal fat with very strong solutions of sodium hydroxide. Heat is needed because the rate of reaction of hydroxide in this case is slow.

The same problems can occur with transition metals. In this lab, you will see that the rate of addition of a molecule called saccharinate is slow at room temperature, but fast at higher temperatures. You will also find that the product $M(Sac)_2$ is much more soluble at elevated temperatures, so simple cooling is required for the isolation of the product.*

---

*This procedure is adapted from that described by Haider, S. Z., Malik, K. M. A., and Ahmed, K. J. 1985. *Inorganic Syntheses* 23:47–51.

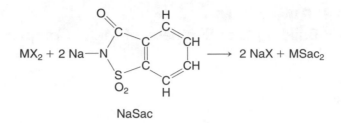

NaSac

The saccharinate in this case is called a *ligand* for the metal. A ligand is a molecule or ion that forms a bond to a metal. Saccharinate is a ligand with a single negative charge, so two saccharinate ligands will complex with a metal with a +2 charge to form a neutral substance. Other ligands may be uncharged (e.g., water or ammonia).

An additional factor for saccharinate complexes is that they form with water molecules also coordinated to the metal. This means the material that forms is a *hydrate*, and the formula is properly written $MSac_2 \cdot 4\,H_2O$. The $\cdot$ symbol indicates that the water molecules are intact, but incorporated in the solid usually as neutral ligands to the metal. In many cases, hydrates can be dehydrated by gentle heating to give the anhydrous salt. Mass calculations for hydrates must be done with the waters included.

CAUTION: This experiment includes the use of sodium saccharinate. Saccharinate is an anion formed by removing a hydrogen ion from saccharin, a substance that is still used today as an artificial sweetener. Both saccharin and saccharinate should be handled carefully because they have been implicated in animal studies as possible carcinogens.

The metal salts in this experiment have been chosen because they do not have significant identified toxicity. However, with all metal salts there are risks associated with high doses. Therefore, handle the solutions carefully.

Procedure

Part I: Formation of Groups

You will be working in groups of six. Each student will begin with 1.00 g of sodium saccharinate but different masses of the metal salt weighing between 0.20 and 0.70 g.

In this laboratory, it is possible to work as if you were blindly following a recipe. This is wrong. Good laboratory technique requires that you keep careful notebook records of what you have done and what you observe in each step. Therefore, prepare your notebook in sections for the preparation of the reaction solution, processing of the reaction mixture to isolate the product, and collection of the product.

For the two experiments in this group, you must report what you observed at each stage. Leave space in your notebook to put observations about the reaction, including colors, whether the mixture is homogeneous or heterogeneous, and the characteristic appearance of the solids.

Part II: Design and Implementation of Synthetic Procedures (Group Work)

Use a centigram balance to measure out the correct amount of each reactant in separate small (25- to 50-mL) beakers. Dissolve each reactant in a small amount of water, between 10 and 20 mL each. The dissolution process may take several minutes.

Once both reactants are dissolved, combine them in one beaker and add a clean stir bar. Rinse the original beakers with 2 to 3 mL of deionized water. Add the rinses to the reactant solutions, place the reaction mixture on a stirrer-hot plate, then start the stirrer. Stir gently to avoid spattering, then turn the heat on to a moderate setting (about halfway on the scale).

Monitor the reaction every few minutes to be certain that there is water left and no large amount of solid has formed. The reaction will come to boiling. At the first indication of solid formation around the edges of the solution, remove the beaker from the hot plate and leave it to cool until it is warm to the touch. At that point, you may put it in an ice bath.

Part III: Practice in Vacuum Filtration (Individual Work)

You will separate the product of the reaction by crystallization and filtration through a Büchner funnel. This is a flat ceramic or plastic device that has many small holes. The holes are too large to keep back most substances, so a piece of filter paper is placed atop the holes.

The filtration is done by putting the funnel on top of a *filter flask*, an Erlenmeyer flask with a side arm near the top. This side arm is hooked up to a vacuum source, usually a vacuum aspirator. This creates a vacuum suction by the rapid flow of water through a small tube. You connect the filter flask to the aspirator using a piece of rubber tubing.

The filtration apparatus must be carefully secured with clamps because it is top-heavy. Also, be careful when removing the vacuum. Do *not* just turn off the aspirator water; this may create a disastrous back-flow of the water into the filter flask. Instead, break the vacuum by disconnecting the rubber tubing from the filter flask.

Begin with a separation of a sample of the compounds salicylic acid and sodium chloride. You will be given a mixture of the two. Obtain about 2 g of the mixture in a small beaker (record the exact mass to at least ±0.01 g).

The sodium chloride is very soluble in water; the salicylic acid is not. Add about 10. mL of water to the mixture. Swirl gently. Turn on the vacuum, then pour the heterogeneous mixture into the filter flask. If any solid comes through the filter paper, then you have not set up the system correctly; redo the experiment. If the solution that comes through (called the filtrate) is clear, then you have done the filtration correctly. Use a wash bottle to rinse any remaining solids from the beaker into the Büchner funnel. Then rinse the solid sitting on the filter paper with a small amount of water.

The flow of air through the filter funnel should remove most of the water from the solid. You may gently stir the filter "cake" with a spatula to aid drying.

Once you are certain that the material is dry, break the vacuum and then collect all of the solids in a clean, dry, and *pre-weighed* beaker. Obtain the mass of salicylic acid, then use this to calculate the percentage by mass of salicylic acid in the mixture.

Part IV: Crystallization and Separation of Product (Individual Work)

After practicing the filtration, you should be ready to isolate the saccharinate salt. If immersion in ice water has *not* produced a significant amount of crystals, then ask your instructor for permission to add seed crystals to the reaction solution. Add *at most* three crystals of the target compound, then wait another 15 min.

The isolation is done by separation with the Büchner funnel. In this case, chill the wash bottle in an ice bath before use. Use this ice-cold water for all rinses.

Dry the product on the Büchner funnel, then collect it and weigh it.

Report

Obtain the expected percentage by mass of the salicylic acid from your instructor and calculate a percentage recovery. Discuss possible sources of error if your percentage recovery is not between 95 and 105%. The acceptable range exceeds 100% recovery. Explain how it is possible to recover more than 100% of the salicylic acid from the mixture of salicylic acid and sodium chloride.

In your report, include the procedure that you used to prepare the metal complex. This should be written in the standard form of a synthetic chemical report. Write the paragraph in the past tense and in the *passive* voice. For example, write "A 2.00-g sample of the compound was weighed" not "I weighed a 2.00-g sample of the compound." Include the mass amounts of all reagents, the volumes of liquids, and the mass amount of the final product. Note the colors observed as well.

You used vacuum filtration to isolate two different substances in this experiment. First, you separated salicylic acid from a mixture of salicylic acid and sodium chloride, then you separated the metal complex from solution. In which case was a physical change involved prior to filtration and in which case was a chemical change involved prior to filtration? Explain. What test(s) could you perform on the collected solids to support your conclusions? (Note: Your instructor may ask you to carry out your tests. If so, include the results of those tests in your report.)

Prepare a table containing the following data from each group member:

- Mass of the metal salt used
- Mass of the product collected
- Theoretical mass of product if all of the metal salt reacts completely
- Theoretical mass of product if all of the sodium saccharinate reacts completely

State the trend between the mass of the product formed and the mass of the metal salt used. For each trial, determine which reactant is the limiting reactant and which reactant is in excess. Then explain how the trend observed between the mass of the product formed and the mass of the metal salt used helps you to identify the limiting reactant. Explain which theoretical mass of product should be used in a calculation of the percentage yield for your trial and then calculate the percentage yield. Tabulate all group results. Finally, summarize what you learned from this experiment regarding reaction stoichiometry and the limiting reactant concept.

EXPERIMENT 2
Application Lab: Gravimetric Analysis of Phosphorus*

Pre-Laboratory Assignment   **Due Before Lab Begins**

NAME: _____

Complete these exercises after reading the experiment but before coming to the laboratory to do it.

1. Answer the following questions about consumer plant foods (or fertilizers) that are labeled 15-30-15.

   (a) What does this labeling system tell you about the composition of the fertilizer?

   (b) Calculate the number of moles of phosphorus (P) in 10.00 g of the 15-30-15 fertilizer.

   (c) If the phosphorus is present in the fertilizer as PO_4^{3-}, how many moles of PO_4^{3-} are present in 10.00 g of the fertilizer?

2. What volume of 0.400 M $MgSO_4$ solution is needed to completely react with the PO_4^{3-} present in 10.00 g of the fertilizer (see your response to Question 1c)?

3. When analyzing the fertilizer for phosphorus content, should you use a stoichiometric amount of $MgSO_4$ solution or should one of the reactants be in excess? Explain.

---

*This experiment follows the procedure outlined in Solomon, S., Lee, A., and Bates, D., *J. Chem. Educ.*, **1993,** 70, 410–412.

4. Why must a base be added in this experiment and why is aqueous ammonia used rather than a strong base such as NaOH?

5. After you filter the precipitate in this experiment, do you expect the filtrate to be acidic, basic, or neutral? Describe what precautions, if any, you must take in handling and disposing of this filtrate.

EXPERIMENT 2
Application Lab: Gravimetric Analysis of Phosphorus

Background

In the skill-building lab, you studied reaction stoichiometry and the limiting reactant concept through the synthesis of a metal complex. In this lab, you will use your knowledge of reaction stoichiometry to again produce a compound, but with the intent of analyzing a fertilizer for its phosphorus content. Specifically, you will perform a gravimetric analysis to determine the percentage by mass of phosphorus in a fertilizer sample.

In gravimetric analyses, the analyte (the substance to be determined) is physically separated from all other components of the sample and the solvent. Precipitation is a common method used to chemically separate the analyte. Because weight is the only measurement needed in gravimetric analyses, the analyte needs to be present in large enough amounts to be weighed in some form—and you must weigh your samples and products as accurately as possible. Therefore, you should use an analytical balance throughout this experiment. Finally, because your goal is to determine the phosphorus content in a fertilizer, you must take care to fully separate the phosphorus from the sample and completely dry the final product.

Most consumer fertilizers are mixtures of water-soluble salts. The phosphorus can be separated out of the mixture by precipitation of magnesium ammonium phosphate:

$$7 H_2O \; (l) + PO_4^{3-} \; (aq) + NH_3 \; (aq) + Mg^{2+} \; (aq) \rightarrow$$
$$MgNH_4PO_4 \cdot 6 H_2O \; (s) + OH^- \; (aq)$$

The solution must be basic in order for the precipitate to form, because the concentration of the phosphate ion decreases in an acid solution due to the formation of hydrogen phosphate:

$$PO_4^{3-} \; (aq) + H_3O^+ \; (aq) \rightarrow HPO_4^{2-} \; (aq) + H_2O \; (l)$$

Ammonia, rather than a strong base like NaOH, is used as the source of the hydroxide ion in the precipitation reaction to avoid the formation of $Mg(OH)_2$.

The precipitate must be dried at room temperature to avoid the loss of the water of hydration. The compound loses all water of hydration at 100 °C.

CAUTION: You will be using an ammonia solution in this experiment. Ammonia is corrosive and can irritate the skin, eyes, and lungs. Avoid direct inhalation of the vapors.

Procedure

Part I: Formation of Groups

You will be working in a group of three people to refine the procedure and to check calculations. Each person will analyze his or her own sample of fertilizer. At least one other group in the class will be analyzing the same fertilizer as your group. Members of all groups analyzing the same fertilizer will pool their results before submitting final reports.

Part II: Determining the Phosphorus Content in a Fertilizer (Individual Work)

The phosphorus in the fertilizer is present in a water-soluble compound, but the fertilizer may also contain some water-insoluble residue. With your group, devise a method to separate the water-soluble from the water-insoluble components of the fertilizer. Once your procedure is approved by your instructor, begin the analysis, each person working with his or her own sample.

Weigh out between 3 and 10 g (to 0.001 g) of fertilizer. Carry out your procedure to separate the soluble and insoluble components.

Other water-soluble compounds are present in the filtrate, along with the phosphorus-containing ions. Therefore, you must now chemically separate the phosphorus from the rest of the mixture. Refer to pre-lab questions 1 and 2 and calculate the volume of $MgSO_4$ solution needed to completely react with the phosphorus present in the sample of fertilizer you measured out. Then add 50% more than the calculated volume of $MgSO_4$ solution. Record your observations. Test the pH of this solution. If necessary, slowly add ammonia solution until the pH is at least 9.0. Record your observations.

Allow this mixture to stand in an ice bath for 30 min. During this time, prepare a filtration flask to collect the precipitate. Weigh the dry filter paper. Consult a chemical handbook or other appropriate reference to determine the properties of magnesium ammonium phosphate hexahydrate. With your group, develop a plan to verify that your precipitate is magnesium ammonium phosphate hexahydrate.

Filter the mixture after it has been in the ice bath for 30 min. Rinse the precipitate with deionized water. Remove the filter paper from the funnel and set it aside to dry in the location designated by your instructor.

Report

When your sample is dry, weigh it and calculate the mass of phosphorus present in the precipitate. Then calculate the mass of P_2O_5 that contains this same mass of phosphorus. Finally, report the percentage of phosphorus as P_2O_5 in the original fertilizer sample. Tabulate the results of all the groups that analyzed the same fertilizer and calculate an average percentage by mass P_2O_5 and a standard deviation.

Group Discussion

Discuss the following questions with members of your group. Incorporate your responses to these questions, along with an explanation of your method and results, into your final report to the fertilizer company that hired you to analyze their product.

1. The large amount of potassium also present in fertilizers could cause the formation of $MgKPO_4 \cdot 6\ H_2O$ in addition to the desired precipitate. When you calculated the percentage by mass of P in the fertilizer, you assumed the precipitate was only $MgNH_4PO_4 \cdot 6\ H_2O$. If some of the solid was actually $MgKPO_4 \cdot 6\ H_2O$, would your calculated percentage be too high, too low, or unaffected? Explain.

2. The fertilizer is a mixture of solids. Can it be a completely homogeneous mixture? Explain. What impact does this have on the precision of your analyses?

3. Fertilizer companies commonly overformulate their fertilizers; that is, the actual content is higher than what is promised on the label. What impact could this have on the results of your analyses? Do your experimental results support the idea that the fertilizer was overformulated? Explain.

Experiment Group **D**

Metal Ions and the Blood: Treat Iron Deficiency . . . and Overload

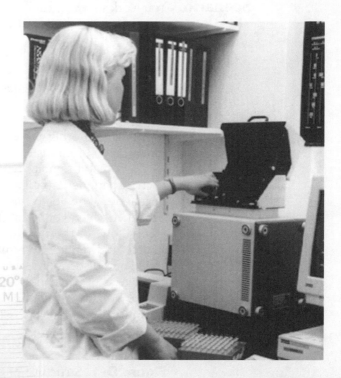

Spectrophotometers are an essential part of many clinical laboratories. (Courtesy of Foss NIR Systems, Inc.)

Purpose
This experiment group introduces you to the way that we can determine the contents of a solution by measuring how it interacts with light. This kind of analysis is called spectrophotometry—the measurement (metering) of the light absorbed (the spectrum) by a solution.

In the case of this set of experiments, we have borrowed material from the department of medicinal chemistry in a college of pharmacy: They use spectrophotometry to determine the amount of iron in blood, part of a diagnosis and treatment. We will not have you work with real blood, but in the second experiment you will carry out procedures that simulate the actual methods.

Schedule of the Labs

Experiment 1: Skill-Building Lab: Spectrophotometry of Dyes

- Prepare a standard solution and a set of diluted solutions of a food dye (individual work).
- Determine the UV-visible spectra of the set of dye solutions (individual work).
- Determine the relative concentration of an unknown solution (group work).
- Combine your solution with other solutions to determine the changes in color and spectrum that result (group work).

Experiment 2: Application Lab: Total Serum Iron Assay

- Construct a calibration curve using standard solutions to match the iron concentration range that might be found in adult males, adult females, and children (group work).

- Measure the total serum iron concentration in a "patient's blood sample" (individual work).
- Treat the patient and measure the serum iron level after treatment (individual work).

Scenario   You work in a clinical laboratory. A common assay used to aid in the diagnosis of diseases affecting the blood (or serum) is for total serum iron. An assay is a test for an amount of a substance. Because of the importance of iron in many biological processes, several health problems arise from levels of iron that are too low or too high.

Iron in food supplements is usually consumed orally in the +2 oxidation state, but in the body it oxidizes to the +3 oxidation state. The Fe^{3+} ions attach to ferritin, a large iron-storage protein (see Figure D-1). Fe^{3+} is released from the ferritin to transferrin, another protein present in the blood plasma, which transports the iron ions to blood forming sites.

Two blood samples have just arrived at the lab. One is from a woman who has been experiencing fatigue, weakness, shortness of breath, low blood pressure, and headaches. The doctor suspects iron-deficiency anemia. Another is from a child who has taken several of his mother's iron supplement pills, thinking they were candy. You receive one of the two samples. Your job is to measure the serum iron level in your sample and report the results to the doctor and pharmacist. They would then prescribe and dispense the necessary treatment to raise the iron level of the woman and lower that of the child. You will then temporarily assume the role of the doctor and decide how to "treat" the patient by either increasing or decreasing the iron concentration in your sample. Next, you assume a pharmacist's role by dispensing the treatment before returning to a technician role to assess the final results.

Figure D-1 Structure of the protein transferrin. Iron binds at several places in both halves of the molecule.

EXPERIMENT 1
Skill-Building Lab: Spectrophotometry of Dyes

Pre-Laboratory Assignment Due Before Lab Begins

NAME: _____

Complete these exercises after reading the experiment but before coming to the laboratory to do it.

1. Consult a color wheel and a graph of the visible spectrum to determine:

    (a) The color complementary to violet.

    (b) The color corresponding to a wavelength of 450 nm.

    (c) The color if all light with wavelengths above 600 nm and below 530 nm were absorbed by a sample.

2. The absorption spectrum of chlorophyll *a* has maxima at 663 and 420 nm. What colors correspond to these wavelengths? What color will be seen reflected by an object with this absorption spectrum?

3. (a) If 5.00 mL of a stock solution is diluted to 10.00 mL, how does the concentration of the dilute solution compare to that of the original stock solution?

    (b) If absorbance is directly related to concentration, how should the absorbance of the dilute soution in part (a) of this question compare to the absorbance of the original stock solution?

4. What is one important safety precaution for this experiment?

EXPERIMENT 1
Skill-Building Lab: Spectrophotometry of Dyes

Background

In this experiment, you will be introduced to the use of UV-visible spectrophotometers, which measure the light absorbed in the ultraviolet and the visible regions of the electromagnetic spectrum.

One type of UV-visible spectrophotometer is the *scanning* spectrophotometer. In it, a grating or prism is physically moved to cause different wavelengths to fall on the single detector. An advantage of scanning is that the optics can be tuned to give high wavelength resolution. However, any scanning system must always include at least one moving part, and, aside from problems of breakage, recording a spectrum may take a considerable length of time because the whole spectrum is sampled one wavelength at a time. The Spectronic-20 spectrophotometers used in many undergraduate labs and in some clinical settings could be used as scanning instruments if you set each new wavelength manually.

The second type of UV-visible spectrophotometer is the *diode-array* spectrophotometer. In it a narrow beam of white light is sent through a sample and then refracted through a prism onto a grid (or array) of hundreds of detectors based on diodes. The detectors are physically arranged so that each position receives only a narrow range of wavelengths (in some instruments 2 nm wide). All of the diodes detect the light simultaneously, so there is a significant decrease in the amount of time it takes to sample many wavelengths. There is only one simple moving part: a shutter. However, the resolution is fixed by the density of the array and can never be improved. The detection of hundreds of different signals requires fast processing and data storage—precisely the strengths of a computer-controlled instrument.

CAUTION: The dyes in this experiment are related to common food dyes, but you should follow appropriate lab procedure at all times. Even when you are using the spectrophotometers, careful handling and safety goggles are essential.

Procedure

There are three parts to this lab. You will prepare a series of solutions of a food dye and obtain their absorbance spectra using the spectrophotometer. Next, you will determine the relative concentration of a diluted solution of your dye. Finally, you will combine your solution with that of another student to get a mixture of dyes, which you will also analyze on the spectrophotometer.

Carefully review the directions for the operation of the spectrophotometer in your laboratory. You will need to know how to obtain and print spectra.

For all of your solutions, including those obtained by dilution, make careful notes of the color and intensity.

Part I: Recording of Standard Spectrum (Individual Work)

When you come to lab, your instructor will assign you a red, yellow, or blue dye. Prepare a stock solution by dissolving four drops of the dye in water in a graduated cylinder to make 100 mL of solution. Mix thoroughly.

When you are ready, use a Mohr pipet to transfer 4.00 mL of your stock solution to a 10-mL volumetric flask. Add deionized (DI) water to the marked line on the flask, cap the flask, and mix well.

Prepare a second sample from 3.00 mL of your stock solution and enough DI water to fill the 10-mL volumetric flask to the line. Mix well.

Prepare a third sample from 2.00 mL of your stock solution and enough DI water to fill the 10-mL volumetric flask to the line. Mix well.

Prepare a fourth sample from 1.00 mL of your stock solution and enough DI water to fill the 10-mL volumetric flask to the line. Mix well. Then transfer a portion of each sample to a different cuvette. Take them to the spectrophotometer, obtain an absorption spectrum for each, and plot the spectra.

Part II: Determining the Relative Concentration of a Dilute Dye Solution (Group Work)

Ask another student to prepare a solution for you by transferring between 1 and 4 mL of *your* stock dye solution into a 10-mL volumetric flask and then diluting it with enough DI water to fill the flask to the marked line. *You* will then transfer a portion of this sample to a cuvette and obtain an absorption spectrum.

Part III: Preparation of Mixtures (Group Work)

You should see that your spectra have a single prominent absorption. Other students with different dyes will have samples that absorb at different wavelengths. Ask one other student to share his or her solution with you. Prepare a solution in a 10-mL volumetric flask using 3.00 mL of your stock solution, 3.00 mL of the other student's stock solution, and enough DI water to fill the flask to the line. Make sure the solution is well mixed. Then transfer a portion to a cuvette and record its spectrum.

Report

Tabulate the wavelengths of the peaks in each of the spectra. Record A values for the peaks. Attach copies of your spectra to your report.

Although you do not have a numerical value for the concentration of the stock dye solution you prepared in the graduated cylinder, you carefully diluted four samples using volumetric glassware. When you transferred 4.00 mL of the stock solution to a 10-mL volumetric flask and diluted with DI water, the new solution became 0.40 times as concentrated as the stock. Therefore, you can plot absorbance versus relative concentration to determine the relationship between these variables. This graph is a calibration curve, which can be used to determine the relative concentration of another sample of this same stock dye.

Using your data from Part I, prepare a graph that sets absorbance on the y-axis and relative concentrations on the x-axis. Use a graphing calculator or spreadsheet to obtain the equation for the best-fit line. Then use this calibration curve to determine how many milliliters of stock are present in the solution of unknown concentration your partner prepared for you in Part II.

Questions to Answer in Your Report

1. For each of the spectra you took, what is the color of the absorbed wavelength? How is that color related to the color you see with your eyes?
2. Beer's law indicates that absorbance is directly proportional to concentration. Are your data consistent with this law? Explain.
3. What happened when you combined your solution with that of another student? How did the spectrum of the mixture compare with the spectra of the individual dyes? Was the intensity of your dye's absorption altered by the presence of the second dye? Explain.

EXPERIMENT 2
Application Lab: Total Serum Iron Assay

Pre-Laboratory Assignment   **Due Before Lab Begins**

NAME: _____

Complete these exercises after reading the experiment but before coming to the laboratory to do it.

1. Since you will be using spectrophotometry to determine the concentration of iron in solutions, you will first generate a calibration curve. What are the dependent and independent variables in this experiment?

2. What is the appropriate absorbance range for spectrophotometry?

3. (a) When using FerroZine® to determine the concentration of iron, how many moles of FerroZine are needed to complex 1.8×10^{-7} mol of iron?

    (b) How many mL of 100 mg/dL FerroZine solution are needed to obtain the number of moles of FerroZine determined in part (a)?

4. (a) How many moles of NH_2OH are present in 0.50 mL of 0.434 M NH_2OH?

    (b) How many moles of Fe^{3+} can be reduced by the NH_2OH you calculated in part (a)?

5. As part of this laboratory, you may have to remove iron from a solution. If you have 20 mL of a solution that contains 43 $\mu g\ dL^{-1}$, how many grams of iron must be removed to reach a concentration of 23 $\mu g\ dL^{-1}$? How many moles of iron is this?

6. As part of this laboratory, you may have to add iron to a solution. If you have 20 mL of a solution that contains 15 μg dL^{-1}, how many grams of iron must be added to reach a concentration of 23 μg dL^{-1}? How many moles of iron is this?

7. What special precautions must you take while handling the hydroxylamine and deferoxamine solutions? Why?

EXPERIMENT 2
Application Lab: Total Serum Iron Assay

Background

The analytical method for this lab is based on a reagent, FerroZine®, that forms an intensely colored solution with iron. FerroZine complexes with Fe^{2+}, forming a magenta colored solution whose intensity is dependent upon the concentration of the complex present. The equation for the reaction, written for the dianion of FerroZine, is:

$$[Fe(H_2O)_6]^{2+} + 3\ FerroZine^{2-} \rightarrow Fe(FerroZine)_3^{4-} + 6\ H_2O$$

The FerroZine is a water soluble salt (with a molar mass of 492.47 g/mol) which forms a -2 anion in solution. Because FerroZine binds Fe^{2+}, all of the iron in solution must be present in the $+2$ state before the concentration is measured. The Fe^{3+} is readily reduced to Fe^{2+} by reaction with excess hydroxylamine:

$$2\ Fe^{3+} + 2\ NH_2OH \rightarrow 2\ Fe^{2+} + N_2 + 2\ H^+ + 2\ H_2O$$

You need to prepare a set of spectrophotometric solutions with different Fe^{2+} concentrations. For each solution, put a fixed amount of standard solution into the cuvette along with a fixed amount of hydroxylamine and FerroZine. This means you will follow a two-step procedure to prepare the spectrophotometric solutions. First, dilute the iron stock solution in a volumetric flask. Then, put a fixed amount into a cuvette, along with a fixed amount of hydroxylamine and FerroZine. You will then measure the absorbance of the solution.

Be sure that there is enough hydroxylamine and FerroZine in each cuvette to reduce and complex all of the iron. To do this, start off by calculating the largest concentration of iron you might have, and then determine how many moles of FerroZine and of hydroxylamine you might need. This will tell you how to make up the cuvettes for the spectrophotometer.

The measurement μg/dL (micrograms per deciliter) is a common concentration unit used in medical analyses. You need to be able to convert between molarity and this unit in order to report your results in the units used in a clinical setting.

CAUTION: You will not be working with real blood samples. Your samples will be solutions of iron(II) ammonium sulfate, $Fe(NH_4)_2(SO_4)_2$, or iron(III) nitrate, $Fe(NO_3)_3$. If you were using real blood samples, you would have to observe the necessary precautions for the safe handling of a blood sample. These include wearing protective gloves, disposing of the gloves and the glassware used to contain the blood sample in a biohazard bag, and washing any spills with bleach.

This lab requires the use of a hydroxylamine solution. It is potentially toxic and a potential mutagen. This lab also involves the use of deferoxamine, a biologically active iron-complexing agent and possible teratogen. The use of proper gloves is advised. Wear appropriate safety goggles. Clean up all spills immediately with excess water. Be careful to avoid ingestion and, as always, wash your hands carefully before leaving the lab.

Procedure

You will have the following materials at your disposal for this lab:

- iron(II) standard solution
- FerroZine solution
- hydroxylamine solution
- solutions containing "normal" levels for iron
- spectrophotometer
- appropriate glassware
- deferoxamine, an iron complexing agent (shown in Figure D-2).

Part I: Formation of Groups

The group in this experiment will simulate a team working in a clinical analytical lab. Three students will share the responsibility for constructing a calibration curve based upon the serum iron levels in "healthy" patients.

Part II: Determining the Response of Healthy Iron Levels in the Assay (Group Work)

The standards will be available in the lab in the form of six "healthy" patient samples. Look on the bottles of "sample sera" available in the lab to get the actual iron concentrations.

These six samples will form your calibration curve for the experiment. Since your unknown patient sample contains either too much iron or too little iron, compared with the acceptable "normal" range for that patient, your calibration curve needs to extend beyond the normal range at both ends. Therefore, you need to prepare two more samples in addition to the six healthy patient samples; one solution that is more concentrated than the highest serum iron level in a healthy patient and one that is more dilute than the lowest serum level in a healthy patient. Use the standard iron(II) solution available to prepare these two solutions in volumetric flasks.

The easiest way to set up the assay is to determine how to take the most concentrated sample and to mix it in a cuvette to give a reasonable value A (between 0.05 and 1.5) in the spectrophotometer, with appropriate amounts of FerroZine and hydroxylamine. Treatment of the other samples with identical dilution methods will fill out the calibration curve.

Part III: Treatment of Unhealthy Iron Levels (Individual Work)

A patient with too much or too little iron must be treated with iron supplements or iron-removal agents. You will simulate this during the lab by adding a supplement or a removal agent to the serum of your patient. Get your patient's serum, and use the method of your group to determine the iron concentration. Then adjust the amount of iron in the serum to bring it to the normal range.

Figure D-2 Molecular structure of deferoxamine, $C_{25}H_{48}N_6O_8$

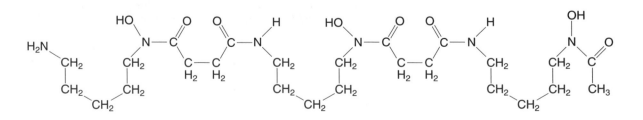

In a clinical setting, compounds that contain iron(II) are used in the treatment of iron-deficiency anemia, which is the condition of too little iron in the blood. If your sample is "anemic," you will mix in a solution of iron(II) sulfate. Calculate how many moles of iron you need to add to the solution, then use the appropriate volume of iron(II) solution to add that amount of iron to your sample. Note that this will dilute your solution by a small amount, but you are aiming for a solution that, when you determine the iron content, is in the normal range for your patient. After mixing in the iron supplement solution, withdraw some of the solution and redetermine the iron concentration.

If you needed to *remove* iron in a clinical setting, you would administer an agent that would bind to the iron and cause it to be excreted. We simulate this by using the agent to bind to some of the iron and to allow the FerroZine to complex less of the iron in the solution—which will show up as a lower iron concentration. Deferoxamine is used in the clinical treatment of acute iron overdoses. Deferoxamine has a molar mass of 656.8 g/mol and binds to iron(III) ions in a one-to-one mole ratio.

To "remove" iron from your solution, you will mix in a solution of deferoxamine. Calculate how many moles of iron you need to remove from the solution, then add the same number of moles of deferoxamine. Note that this will dilute your solution by a small amount, but you are aiming for a solution that, when you determine the iron content, is in the normal range for your patient. After mixing in the deferoxamine solution, withdraw some of the solution and redetermine the iron concentration.

Results and Conclusions

Organize your experimental results in an appropriate manner. Include graphs if any were generated.

Your conclusion should include a summary of the experimental technique you and your group developed in order to measure the serum iron level in the patient's blood. What are normal serum levels for adult males, females, and children? What were the results of your patient's sample? How did you treat the patient? Was the treatment successful? Explain how you know.

Questions to Answer in Your Report

1. Iron(II) sulfate and iron(II) fumarate are often used in iron supplements for the treatment of iron-deficiency anemia. Why do you think these compounds are used?

2. The body retains only 10 to 15% of the iron consumed. A prenatal multivitamin contains 45 mg of iron per tablet. Assuming maximum retention (15%) and a volume of serum of 8.0 L, determine the concentration, in micrograms per deciliter, of iron after ingestion of the vitamin.

Experiment Group **E**

Analysis of Complex Solutions: Bloodwork Quick with a Finger Stick

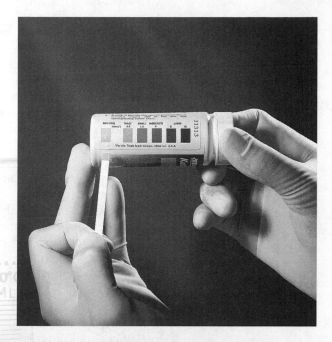

Measurement with a test strip is an important way to quickly assess a patient's health. (Saturn Stills/Science Photo Library/Photo Researchers.)

Purpose Chemists often use the way that light interacts with matter, also known as spectrophotometry, to determine how chemical substances interact with each other. This is especially useful in studying solutions, including solutions in which two or more substances form a complex with each other. This experiment group will develop your skills in analyzing a solution using UV and visible light, and will show you how the analysis of a complex of a dye with a blood protein permits the determination of important health-related parameters.

Schedule of the Labs

Experiment 1: Skill-Building Lab: Stoichiometry of a Metal–Ligand Complex
- Microscale determination of molar ratios (group work).
- Preparation of solutions for mole ratio analysis (individual work).
- Spectrophotometric determination of the formula of a complex (group work).

Experiment 2: Application Lab: Determination of Serum Albumin
- Spectrophotometric study of the binding of bromocresol green and albumin (group work).
- Design of a finger-stick test (group work).
- Measurement of serum albumin (individual work).

Scenario Any organism with a circulatory system has the challenge of forcing blood through tubes known as arteries and veins. The heart applies the pressure needed for blood to flow, generating a mechanical pressure known as the blood pressure.

The arteries and veins, especially when very small, have walls that are permeable to water and other small uncharged solutes. But ions and larger molecules cannot pass through the walls of the blood vessels. The blood pressure could, if left unchecked, force most of the water from the blood, much as water can seep through a wall with cracks.

This pressure is mechanical and can be compared to squeezing water through a fine filter. The only way to counteract the blood's mechanical pressure is osmotic pressure, which describes the tendency of water to be retained by a solution. Osmotic pressure holds back the water that would be squeezed out by blood pressure.

Maintaining osmotic pressure requires a solute that will *not* pass out of the blood vessels. In mammalian blood, the osmotic pressure is regulated by a large molecule called *albumin*. Proper concentrations of albumin are found in healthy individuals, but many diseases can cause albumin to be lost via the urine, affecting the entire circulatory system.

One of the first analyses done when a patient is admitted to a hospital is the determination of the serum (blood) albumin level. This is done on an analyzer that uses the exact same method you will use in this experiment group. The serum is diluted into a solution containing an acid–base buffer and an excess of the dye bromocresol green (BCG). BCG is normally yellow under these conditions, but it forms an intensely green-colored complex with albumin. The green albumin–BCG complex is the basis for this experiment.

Many analytical methods rely on a less precise but much quicker method of simple finger-stick analysis. A drop of blood is placed on a small reagent strip and the chemical substances on the strip change color, depending on the amount of material in the blood. Comparison of the color of the strip with a standard color strip provides a rough value for the concentration of the material in the blood.

In this group, you will carry out both finger-stick and more precise determinations of the albumin concentration, simulating the professionals who must obtain quick results prior to sending certain samples to a lab.

EXPERIMENT 1
Skill-Building Lab: Stoichiometry of a Metal–Ligand Complex

Pre-Laboratory Assignment   **Due Before Lab Begins**

NAME: _____

Complete these exercises after reading the experiment but before coming to the laboratory to do it.

1. A solution is made by dissolving 1.00 g of the ligand bipyridine ($C_{10}H_8N_2$) in enough water to make 1.00 L of solution. Determine the concentration of bipyridine in the following solutions:

    (a) Take 10.0 mL of the original solution and add enough water to make 100. mL of solution.

    (b) Take 0.40 mL of the original solution and mix in a cuvette with 0.50 mL of hydroxylamine and 1.40 mL of iron solution.

    (c) Mix 1.00 mL of the solution from part (a) with 1.20 mL of hydroxylamine solution and 0.40 mL of iron solution.

2. In this lab, you will also carry out dilutions using a dropper to get qualitative information about metal/ligand ratios. Determine the approximate answers to these questions:

    (a) Put two drops of hydroxylamine solution in a 0.50-mL well. Then add eight drops of 1.00 g L^{-1} bipyridine and two drops of 5 mg L^{-1} iron. How many moles of bipyridine and how many moles of iron are in the well? There are about 20 drops to 1 mL.

    (b) If you want to repeat the mixture from part (a) on a larger scale, describe how many milliliters of hydroxylamine solution, how many milliliters of bipyridine solution, and how many milliliters of iron solution you need to make 2.50 mL of total solution volume.

3. List two different safety precautions that are important in this laboratory.

EXPERIMENT 1
Skill-Building Lab: Stoichiometry of a Metal–Ligand Complex

Background

A study of the absorption of electromagnetic radiation by a sample is one of the best ways chemists have to determine the chemical contents of a solution without an extensive process of reaction analysis that usually causes the destruction of the sample. UV-visible spectrophotometry is also called electronic spectrophotometry, because absorption of UV and visible light often causes a redistribution of the electrons in the absorbing species.

In this experiment group, you will study the nondestructive examination of a set of solutions to determine their chemical contents. You examine how spectrophotometry can be used to answer the question, "How much of a certain substance is in the solution?" You will also work on the problem of determining *what* is in the solution.

Determination of Mole Ratios in Complexation: A Graphical Method*

If any two substances A and B are mixed in solution, they may in principle react to make one or more new substances. In many cases, this is a simple addition reaction in which one or more moles of substance B add to a mole of A to give a complex:

$$A + nB \rightarrow AB_n$$

The species AB_n is referred to as a complex of A and B. A common type of complexation occurs when A is a metal ion that reacts with other small molecules or ions (B) to form a metal coordination complex. In those cases, the term *coordination* refers to the connection of B and A, so the A—B bond is sometimes called a *coordinate bond*. In such a complex, B is referred to as a *ligand* for A. This experiment uses an elegant method to determine when two substances are present in the right ratio for formation of a complex.

The amount of AB_n formed depends on the amounts of A and B in the mixture and on the value of n. Let's say that n is 4. This is the case, for example, in the formation of the complex $[Cu(NH_3)_4]^{2+}$ when a solution of a copper(II) salt is mixed with concentrated ammonia solution:

$$Cu^{2+}\,(aq) + 4\,NH_3\,(aq) \rightarrow [Cu(NH_3)_4]^{2+}\,(aq)$$

In this case, ammonia acts as a ligand.

If we prepare a solution that contains 0.0020 mol of Cu^{2+}, the amount of complex that forms depends on which reactant is the limiting reactant. If there is less than 0.0080 mol of ammonia in the solution, then the ammonia is the limiting reactant and, regardless of the amount of copper, we will have only as much complex as the supply of ammonia permits.

If we have more than 0.0080 mol of ammonia, copper is the limiting reactant. Then we would form 0.0020 mol of $[Cu(NH_3)_4]^{2+}$ complex.

---

*This method follows that outlined in Po, H. N., and Huang, K. S.-C. *J. Chem. Educ.* **1995**, *72*, 62–63.

You may be familiar with using spectrophotometry to determine the amount of a substance in a solution. But did you ever wonder where the recipe for mixing comes from? It requires that someone carefully determine the relative values of the two reactants, so that every solution that is analyzed has enough ligand. The point of today's exercise is to for you to determine n for a series of ligands that complex with Fe^{2+}.

Two Important Points

Protection of Iron(II) from Oxygen

We will be working with the Fe^{2+} ion in this experiment. However, this ion reacts rapidly with air to give Fe^{3+} in acidic or even neutral solutions:

$$4\ Fe^{2+}\ (aq) + O_2\ (g) + 4\ H^+\ (aq) \rightarrow 4\ Fe^{3+}\ (aq) + 2\ H_2O\ (l)$$

Fortunately, if you do not splash, stir, or shake a solution rapidly, then oxygen is slow to get into the solution. To correct for the small amount of Fe^{3+} that forms during normal handling, we always put some of the reducing agent hydroxylamine, NH_2OH, in the solution to keep all the iron in the +2 state.

Determination of a Ligand-to-Metal Ratio

We determine n by the addition of increasing amounts of ligand to solutions that contain less and less metal. In the first solutions, there is not enough ligand: The ligand is the limiting reactant. In the later solutions, there is not enough metal. The procedure gives rise to a "peak" solution, in which the ligand-to-metal ratio is just right. That solution has a distinct color compared to those before and after it in the series. We then find the ratio of ligand to metal that went into this solution.

Table E-1 gives an idea of how this method works. The ligand-to-metal ratio n is 2. We start with 1.0×10^{-6} mol (or 1.0 micromole, 1.0 μmol) of metal, and 0.25×10^{-6} mol (0.25 μmol) of ligand. In each successive solution, we increase the volume of the ligand solution by 0.20 mL and decrease the volume of the metal solution by the same amount. As a result, the number of moles of complex increases, then begins to decrease once we pass the ideal 2:1 ratio. This occurs somewhere between sample C (when the ligand-to-metal ratio is 1.25:1) and sample D (when the ligand-to-metal ratio is 2.5:1).

TABLE E-1

Solution	Volume of metal (0.0010 M)	Moles of metal	Volume of ligand (0.00125 M)	Moles of ligand	Moles of complex
A	1.00 mL	1.00×10^{-6}	0.20 mL	0.25×10^{-6}	0.12×10^{-6}
B	0.80 mL	0.80×10^{-6}	0.40 mL	0.50×10^{-6}	0.25×10^{-6}
C	0.60 mL	0.60×10^{-6}	0.60 mL	0.75×10^{-6}	0.38×10^{-6}
D	0.40 mL	0.40×10^{-6}	0.80 mL	1.00×10^{-6}	0.40×10^{-6}
E	0.20 mL	0.20×10^{-6}	1.00 mL	1.25×10^{-6}	0.20×10^{-6}

We can demonstrate this method by a simple experiment with drops. We systematically vary the ligand-to-metal ratio by mixing different numbers of drops of each solution, along with any other necessary solution (e.g., hydroxylamine) in the wells

Figure E-1

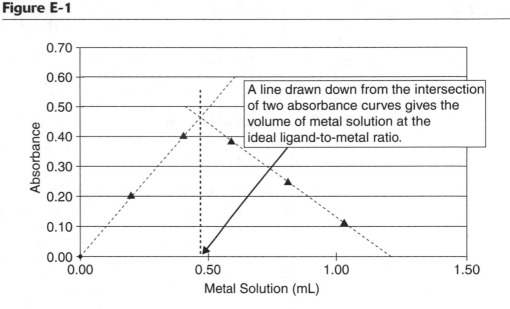

A line drawn down from the intersection of two absorbance curves gives the volume of metal solution at the ideal ligand-to-metal ratio.

of a tissue culture plate. Observation may show a continuous increase or decrease in color, which means that we have the same reactant in excess in all cases. But if we see a "break" in the color, then we have, somewhere in the series, a solution with the ideal ligand-to-metal ratio. For example, in the experiment shown in Table E-1, the deepest color would occur in solution D.

To get an exact value for the ratio, we use the spectrophotometer. The instrument should be set to record the optimum wavelength for the complex (consult with your instructor about how to do this). We prepare a series of mixtures that have the same ligand-to-metal ratios that gave the break in the color in the test done with drops. These mixtures may be prepared in volumetric glassware, followed by transfer to cuvettes, or they can be prepared directly in the cuvettes.

The absorbance values A for the five solutions are now recorded at the optimum wavelength. These results are graphed with milliliters of metal solution on the x-axis and A on the y-axis. A properly done graph will look something like the one in Figure E-1. Lines are drawn as shown along the two legs of the graph, and their intersection is noted. This point lies above the spot on the x-axis where the ligand-to-metal ratio is equal to the correct stoichiometric value. From this volume of metal and the corresponding value of the volume of ligand, we can determine the moles of ligand and the moles of metal in the complex. We have determined n.

 CAUTION: This laboratory requires the use of hydroxylamine solutions. These are potentially toxic and are potential mutagens. Use of proper gloves is advised. Clean up all spills immediately with excess water.

The ligands have a high affinity for iron. Be careful to avoid ingestion; as always, wash your hands carefully before leaving the lab.

Procedure

There are three parts to this lab. After forming groups and dividing up the ligands, you will use droppers and multiwell tissue culture plates to determine an approximate value for the stoichiometric ratio of volumes of the ligand and the metal. Then you will repeat this procedure, with volumetric glassware and a spectrophotometer.

Part I: Formation of Groups

You will work in groups of three today. Each student works with a different ligand. Some of the ligands will require more dilution than others.

Part II: Microscale Determination of Ligand-to-Metal Ratio

You will have stock solutions for the reaction of iron(II) with three different ligands: FerroZine®, terpyridine, and phenanthroline. Before you do any work with the solutions, record the concentrations that are listed on the bottles. Obtain 50 mL of the iron and the ligand solutions and 10 mL of hydroxylamine solution in separate clean, dry beakers.

All three ligands (Figure E-2) complex Fe^{2+} ion through nitrogen atoms. One, two, three, or even more of the ligands may complex with the metal at one time. The first step is to determine how much metal and ligand are needed to provide a good plot. You do this with a tissue culture plate that has wells with a volume of 0.5 mL or less.

Take the plate and align it so that, from left to right, you have at least five wells. Add three drops of hydroxylamine to each. Then add 2, 4, 6, 8, and 10 drops of the iron solution to the wells, starting at the left and moving right. Finally, add 10, 8, 6, 4, and 2 drops of the ligand solution, again moving from left to right. Gently swirl the plate to mix the solutions, and put the plate down on a white piece of paper.

Figure E-2 Ligands for analysis of iron

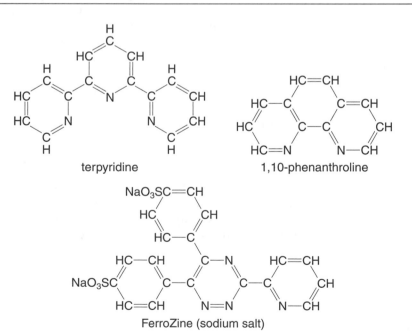

terpyridine   1,10-phenanthroline

FerroZine (sodium salt)

Observe the solutions, making careful note of the color and the intensity of the color. Do you see a break in the color? If you do, then you have a range of relative amounts that includes the stoichiometric amount. You might not see a break in the color. That means you do not have the optimum metal-to-ligand ratio in any solution.

If you have no break in the color, then you need to think about which reactant to adjust. If you find that the color increases directly with the amount of ligand, then the *ligand is the limiting reactant* in all cases. You must then *dilute the iron*. If, on the other hand, the solutions become steadily darker with increasing metal ion, then the *iron is the limiting reactant* and you need to *dilute the ligand*.

You began with solutions that cover a 1:5 to 5:1 ratio of volumes of metal to ligand. If it is necessary to dilute the metal or the ligand to bring it into the range of a proper plot, then take the appropriate solution and dilute it by a factor of five. This is done by measuring 20.00 mL of the original solution into a 100.0-mL volumetric flask and adding enough water to bring the total solution volume to the mark on the volumetric flask. Do not forget to mix well.

If a dilution was necessary, then after you have diluted the solution you should carry out the tissue plate analysis again. This time you should see that you have a well in the center of the range that is the darkest. If not, then try a 10:1 dilution of the original solution.

Part III: Mole Ratio Analysis

Your group now has a solution of the metal and each of the ligands that are appropriate for spectrophotometric analysis. The samples for mole ratio analysis are prepared in the same ratios as the "successful" tissue culture plate experiment. You will have the diluted solutions already (if dilution was needed). Now, prepare five cuvettes with the same volume ratios of the metal solution to the ligand solution. In all cases, there should be a consistent amount of hydroxylamine in each cuvette, and then a consistent total volume.

This can be planned by filling in the worksheet on the next page.

Wavelengths for maximum absorption of iron(II) complexes

Fe^{2+}–FerroZine	562 nm
Fe^{2+}–phenanthroline	510 nm
Fe^{2+}–terpyridine	552 nm

Plotting and Analysis of Spectra

In your group, compare your results and determine if you have the same trends as each other. You may want to work together to convert the volume measurements you have made into the mole amounts you will need for your report.

Report

Your report should discuss how the two methods, tissue culture plate and spectrophotometric, both gave the information needed to determine the correct ligand-to-metal ratio. Present a figure showing the wells in the tissue culture plate as a function of the number of drops of the metal complex and the ligand (after dilution, if that was needed). Label them carefully by color.

Worksheet to Guide the Preparation of Solutions

IMPORTANT: This is to assist you in planning. All data used in the actual experiment should be recorded in your laboratory notebook.

Part I: Tissue culture plate analysis: Fill in the number of drops of each component in the different wells of the tissue culture plate.

	Solution				
	A	B	C	D	E
Drops NH$_2$OH (should be constant)					
Drops Fe^{2+}					
Drops ligand					
Total number of drops (should be constant)					

Part II: Determine the conversion factor between tissue culture plate and cuvette.

Cuvette volume (mL):	$\dfrac{\text{Cuvette volume}}{\text{total numbers of drops}}$:	mL drop^{-1}

Part III: Spectrophotometric analysis with cuvettes. Multiply each of the values in Part I of the worksheet by the conversion factor in Part II to get the volume in milliliters to use in the cuvette.

	Solution				
	A	B	C	D	E
Milliliters of NH$_2$OH (should be constant)					
Milliliters of Fe^{2+}					
Milliliters of ligand					

For the spectrophotometric experiments, tabulate the volume and concentration of both the metal and the ligand in each of the solutions. Also include a value for A in each of the solutions. A graph presenting volume of metal solution and A, similar to Figure E-1, should be prepared, showing how the two lines converge to the ideal value. Finally, calculate the value of n from the ideal mole amounts of the metal and the ligand.

EXPERIMENT 2
Application Lab: Determination of Serum Albumin

Pre-Laboratory Assignment   **Due Before Lab Begins**

NAME: _____

Complete these exercises after reading the experiment but before coming to the laboratory to do it.

1. During this laboratory, you will have to assess the "health" of a "patient" based on a sample of "blood." If you determine that this sample has an albumin concentration of 60 g L^{-1}, is the patient healthy or not? Explain your answer.

2. You mix 0.20 mL of a solution containing 6 g L^{-1} albumin with 2.80 mL of BCG reagent. Assuming the volumes are additive, what is the concentration of albumin in the resulting solution?

3. Beer's law is assumed to be valid in this experiment. Will Beer's law apply if we choose to use concentrations in g L^{-1}? Explain your answer.

4. Since you are not working with human blood in this lab, is there any reason to worry about the safe handling of solutions in this lab? Explain your answer.

EXPERIMENT 2
Application Lab: Determination of Serum Albumin

Background

With the experience of the skill-building lab, you are ready to carry out the analysis of the serum protein albumin in blood. You will again do this in tissue culture plates and using a spectrophotometer. The method is based on the reaction of albumin with a dye, bromocresol green. At the pH of the experiment, bromocresol green is yellow, but when albumin is present, BCG forms a bright green complex with albumin. This reaction can be used at either the tissue culture plate or the spectrophotometric scale.

Table E-2 indicates the expected range for albumin levels in the blood of healthy adult humans. Your method will permit the determination of albumin concentrations. In practice, other serum proteins, especially globulin, react with BCG. That will not be a problem in your experiments here, but in the clinic it is the reason why very precise albumin concentrations require more tedious and expensive methods. Nevertheless, initial screening for abnormal levels of albumin is effectively done with BCG, and then additional tests are done where required.

TABLE **E-2**

Normal Ranges for Albumin in the Blood		
Protein	Normal levels	Possible values
Albumin	35–55 g L^{-1}	20–70 g L^{-1}

You and your group will develop two methods to determine albumin. One will be done with just a couple of drops of sample, similar to the finger-stick methods you may be familiar with. To complement this, you will also find out how to determine albumin levels quantitatively. At the end of the lab period, you will get albumin from a series of "patients," and use the approximate method to determine which require additional, quantitative, study.

Notes:

- You are not working with real human blood or even human albumin. All solutions are prepared from purified equine (horse) proteins. But, since this lab simulates the development of a method for human samples, you should practice good laboratory hygiene and do all manipulations using laboratory gloves. Gloves will be available for your use.

- The clinical analysis is done using very small syringes to measure serum amounts, typically 50 μL. Such syringes are not available for us, but small, 1.0-mL Mohr pipets are. To make it possible for you to measure solutions conveniently, all samples have been diluted by a factor of 40 relative to standard amounts. Thus, the stock solution we provide (2.50 g L^{-1}) will give a response equal to a patient with 2.5 g L^{-1} × 40 = 100. g L^{-1}. If you want to measure a sample equivalent to 20 g L^{-1} in blood, you want to make a solution that has 20/40 = 0.50 g L^{-1} from the stock solution we provide.

CAUTION: You will not be working with real blood samples. Your samples will be solutions of equine (horse) proteins. If you were using real blood samples, you would have to observe the necessary precautions for the safe handling of a blood sample. These include wearing latex gloves, disposal of the glassware used to contain the blood sample and the latex gloves in a biohazard bag after analysis is complete, and washing any spills with bleach.

In this lab, chemical-resistant gloves will be available. Their use is advised, because the BCG solution that you will use is a concentrated buffer solution.

Procedure

Part I: Formation of Groups

Groups of four will be responsible for testing the albumin–BCG reaction, and each group member will prepare two solutions for the common calibration curve. Each member will have a unique set of unknowns to test.

Part II: Characterize the Albumin–BCG Reaction and Prepare a Calibration Curve (Group Work)

Each group will get an analysis kit containing the standard solutions of albumin (2.5 g L^{-1}) and bromocresol green. You will also find a solution of "standard" NaCl, which should be used for *all* protein dilutions and to fill out the volume required for the cuvettes.

It is important that the group members make observations together on albumin. Practice together adding a few drops of the albumin solution to about 1 mL of the BCG reagent. Watch carefully for differences in the response of the reagent. Check it with the spectrophotometer at 1 min and at 5 min after it is prepared. If you cannot record the entire UV-visible spectrum efficiently, then you may select the best wavelength by referring to the spectra in Figure E-3.

Once you have an idea of the response of the protein to the reagent, use standard solutions to design and test a quantitative method. First, make sure that you will always have enough BCG reagent by carrying out a tissue culture plate test. Add varying amounts of albumin to a *consistent* set of solutions of BCG and NaCl. In this case, albumin should *always* be the limiting reactant, therefore you never see any break in the color.

Figure E-3 UV-visible spectra of solutions of albumin with BCG reagent

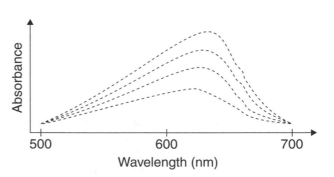

Your work will conclude by preparing eight solutions. These should cover the range of possible values in Table E-2 for the concentration of serum albumin. Consult the list of possible ranges (not just healthy ranges) and prepare a 10.00-mL volumetric flask for each possible value of albumin. Use the absorbance of these solutions to get the calibration curve and to prepare the reference sheet for the finger-stick analysis.

Part III: Preparation of a Finger-Stick Analysis (Individual Work)

In clinical settings, it is often important to get a variety of semiquantitative measurements in a very short time. This is done by wetting a piece of pretreated paper with a sample of blood. The change in the color of the paper indicates the approximate amount of the albumin (or other substance) that is present.

Simulate this using the tissue culture plates in the laboratory. Prepare eight wells with the same amount of BCG solution in each. Add the same amount (say, two or four drops) of one of the group's albumin standards to each and allow the color to develop. These make up your reference sheet for the next step.

Each individual will receive an assignment to test five different patients. The samples can be obtained from the eye dropper bottles. Transfer several drops of the samples for your assigned patients into a set of vials. Use the tissue culture plate test to determine the concentration of albumin in each patient. Compare those with the reference wells. Try to estimate the albumin to within 5 g L^{-1}.

Part IV: Spectrophotometric Determination of Serum Albumin (Individual Work)

Two of the five samples you test will be for patients with either high or low albumin levels. These must be tested further before medical treatment is begun! Return to the patient and get some more sample, but not more than 1 mL. Use this, along with the calibration curve and the spectrophotometer, to determine the concentration of albumin to within 2 g L^{-1}.

Calculations and Report

During the Laboratory

Carry out a quick analysis of the unknown you are studying. Determine the presence of low, normal, or high concentrations of albumin in each of the five patients, and then report the concentration in grams per liter of albumin. Turn in the "stat" report form to your instructor to indicate whether the patient is in danger because of an abnormal level of albumin.

After the Laboratory

Your report should start with a one-paragraph standard procedure that can be used to analyze samples in the future. Next, include a section discussing any restrictions on the method: what ranges of albumin will *not* work with this method, and any information about how to handle the sample. Finally, present one paragraph indicating why this method may be very helpful in rapid determination of albumin and another paragraph indicating why it might *not* be helpful.

Experiment Group F

Acid–Base Titration and the Global Carbon Cycle: Predict Effects of Rising Levels of CO_2

Collecting plant material is a first step in many ecological investigations. (D. Thomas/Visuals Unlimited.)

Purpose

This set of experiments will give you experience in acid–base titration and its use in the analysis of mixtures. In the first week, you will become familiar with acid–base titrations and back titrations as you analyze different antacids. In the second week, you will use these techniques in the analysis of the amount of CO_2 produced by decomposing leaves.

The scenario for this module is taken from the actual methods employed by field ecologists when they need to determine how rapidly biomass on the forest floor—especially leaves—is decomposing. This is an important factor in the balance of carbon dioxide in the atmosphere, and this in turn may be important in climate control.

Schedule of the Labs

Experiment 1: Skill-Building Lab: Analysis of an Antacid

- Analyze different antacids by neutralization with an excess of a solution of HCl and titration of the unreacted acid with NaOH (individual work).
- Determine the neutralizing capacity of an antacid as moles of hydrogen ion (n_{H^+}) per gram of tablet. This is the ratio n_{H^+}/g (individual work).
- Compare the effectiveness of different antacids (group work).
- Set up two leaf compost chambers and a control jar for the second week of the experiment (individual work).

Experiment 2: Application Lab: Tree Leaves and the Global Carbon Cycle

- Test the solubility of metal cations to determine which is best for the selective precipitation of carbonate in the presence of hydroxide (group work).
- Test the method for the determination of NaOH in a $NaOH/Na_2CO_3$ mixture. Add excess metal cation solution to the mixture, then titrate (individual work).
- Use the procedure just tested to determine the number of moles of unreacted NaOH in the experimental setups (leaf compost jars and control jar) (individual work).
- Calculate the number of grams of CO_2 produced per gram of dry leaf per day under varied conditions (individual work).

Scenario   You have a summer internship! A biologist wants to use your knowledge of chemistry to help predict the effects of rising atmospheric CO_2. She tells you that the CO_2 concentration in the atmosphere is rising because of the burning of petroleum, coal, and wood. She says that predicting the effects of an atmosphere with a high CO_2 concentration on society is important. It is possible to imagine, she says, negative effects such as climate warming leading to sea-level rise and coastal flooding. But, she continues, it is also possible to imagine positive effects such as improved plant growth and more land in the north available for agriculture.

The biologist shows you the equation

$$C_n(H_2O)_m + n\,O_2 \rightarrow n\,CO_2 + m\,H_2O + \text{energy}$$

She grows plants in chambers containing elevated partial pressures of CO_2 in order to study the right-to-left reaction. She suggests that you study the left-to-right reaction, which gives the amount of CO_2 returned to the atmosphere by decomposing leaves. She says there is great interest in whether some of the carbon fixed by plants fails to decompose and is thus kept out of the atmosphere. And with that, she goes back to her own work.

But what exactly should you do? To decide, you spend some time in the library reading about the effect that fossil fuel use has on the carbon cycle (see Figure F-1). You learn two things. First, almost 90% of the carbon returned to the atmosphere is from the respiration of decomposers (bacteria and fungi) breaking down dead leaves. Second, the burning fossil fuels release other substances in addition to CO_2. One of these is nitrate, which, like CO_2, is a plant nutrient.

Figure F-1 The global carbon cycle; figures in parentheses refer to billions of tons (after J. E. Ferguson, *Inorganic Chemistry and the Earth;* Pergamon Press: London, 1982).

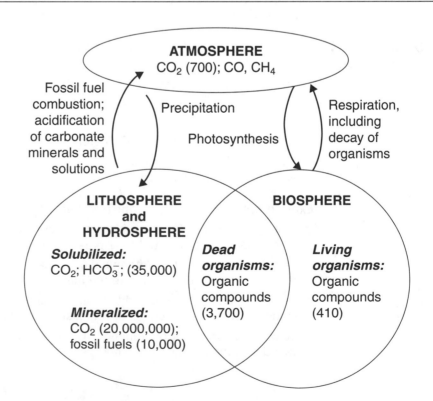

You decide to do two experiments that will answer the biologist's question and a question that formed in your mind as you read in the library. First, the biologist asked whether some of the carbon fixed by plants fails to decompose. You don't have time to study wood decomposition, and you can't visit peat bogs where carbon probably is accumulating. So you decide to use leaves and ask whether they could decompose in a year under favorable conditions. You reason that if the number of days it would take leaves to decompose under favorable lab conditions is more than a year, then carbon will accumulate and there will be less in the atmosphere.

Your library reading made you curious about a second question that your biologist supervisor didn't say anything about. The nitrate that is released from burning fossil fuels is an important plant nutrient. Could nitrate increase the decomposition rate? If so, that would be a reason to expect rapid decomposition of leaves and little accumulation of carbon in dead plant material.

EXPERIMENT 1
Skill-Building Lab: Analysis of an Antacid

Pre-Laboratory Assignment   **Due Before Lab Begins**

NAME: _____

Complete these exercises after reading the experiment but before coming to the laboratory to do it.

1. Why do you pre-rinse the buret with the solution you will be dispensing from it?

2. What do you think will happen on a molecular level when you add the antacid to the HCl solution? Why?

3. What purpose does the indicator serve?

4. After adding the antacid to the HCl in this experiment, do you expect the resulting solution to be acidic, basic, or neutral? Explain.

5. What precautions should you take while heating the antacid–acid mixture? Why?

EXPERIMENT 1
Skill-Building Lab: Analysis of an Antacid

Background

Before you can analyze CO_2 from dead leaves, you must first develop and test a method of analysis. You will test this method on an antacid tablet, which will respond similarly to your method as will the products of your decomposing dead leaves.

A popular ad shown several years ago pictured a person in obvious discomfort saying "I can't believe I ate the whole thing." If you have ever found yourself in a similar situation, do you respond as the person in the ad did by popping an antacid? What does an antacid do to relieve your discomfort? The pH in your stomach is normally 2 to 3. When your stomach pH level drops, you may take an antacid to neutralize the excess acid and return the pH level to normal (though not all the way to a pH of 7— you don't want to neutralize *all* of the acid).

If you use an antacid, how did you decide which brand to buy? In this experiment, you will compare different brands of antacids. Are all antacids equally effective in neutralizing acid? This is what you will determine in this experiment.

To determine how much acid an antacid tablet can neutralize, you will dissolve the antacid in *excess* acid and then titrate the unreacted acid with a standard NaOH solution. You will know how much acid you started with and how much of it reacted with the NaOH, so you will be able to determine how much reacted with the antacid.

CAUTION: You will be working with acids and bases in this experiment. These solutions are corrosive. Wash all small spills (drops) with excess water. Notify your instructor in the event of a large spill (several milliliters). Always wear appropriate safety goggles and wash your hands thoroughly before leaving the lab.

Procedure

Part I: Formation of Groups

We recommend working with no more than three people per group. Different groups should analyze different antacids, in order to compare brands, and each group member should perform a minimum of two analyses so that the group has at least six results for the same antacid. Then each group should reconvene to discuss results.

Part II: Acid–Base Titration of Antacid (Individual Work)

Obtain two or more burets for the group, one for use with HCl and the others for NaOH. Remember to make sure the burets are clean, to rinse them with deionized water, and to pre-rinse with the solution they will each contain. Label the burets so you know which are for the NaOH and which is for the HCl.

Determine the mass of each of your antacid tablets. When you obtain the HCl and NaOH, record their exact molarity as they appear on the stock bottles.

Dispense no more than 25 mL of HCl solution into each of two clean Erlenmeyer flasks. Record the exact volume of HCl used in each. Then add one of the antacid tablets to each flask and heat gently for 1 or 2 min to help to dissolve the tablets. Bring each solution to a boil to dispel any undissolved CO_2. Cool. Some solid materials, used to bind the antacid tablet together, may remain.

Add two or three drops of an indicator to each of your antacid mixtures. See your instructor if your solutions are not the color you expect, keeping in mind that you want an excess of acid present in each mixture.

Titrate each antacid mixture with the NaOH solution until you have reached the end point of the indicator used.

Calculations

1. For each tablet, calculate the moles of hydrogen ion neutralized by the antacid per gram of antacid tablet (n_{H+}/g).

2. Determine the group average n_{H+}/g of antacid and the standard deviation. If the group decides to eliminate any values from the group average, justify the decision to do so.

3. Calculate the average mass of the antacid tablets.

4. Calculate the cost of the antacid per gram in cents per gram.

5. Write the balanced chemical equation for the reaction between HCl and the active ingredient in your antacid tablet. Then, using your experimental data, calculate the mass of the active ingredient per tablet in your antacid.

Report

Compare the precision of your own trials with each other and with that of the other members of your group. Account for differences.

How does the calculated mass of the active ingredient compare with the information on the label? Account for differences.

This laboratory uses a *back titration*. How does this method differ from the direct addition of HCl to the antacid? Why do you think a back titration is used here?

Report to the entire class the ingredients in your antacid, the average n_{H+}/g of a tablet, and the cost per gram. After all the class data are tabulated, discuss which antacid you would buy and why. Are there considerations besides cost and neutralizing ability? Include a summary of your group discussion and conclusions in your report.

Preparation for Week 2 Experiment

Set up your two leaf compost jars and a control jar and leave all three jars in the same location until next week.

1. Obtain two sets of three or four leaves each (use the same type of leaf in both jars—for example, maple, oak), three airtight jars, three small beakers, and a 20-mL pipet.

2. Weigh each set of the dry leaves separately.

3. Using a pipet, measure 20.0 mL of the approximately 1.00 M NaOH that you used in the antacid experiment into each of the three small beakers. Record the exact molarity of NaOH solution.

4. Place one of the small beakers containing NaOH into one of the jars and seal the jar.

5. Place one set of leaves into one of the other jars. Spray some water on the leaves, then place the second small beaker with NaOH on top of the leaves. Be careful to avoid direct contact between the leaves and the NaOH solution. Seal the jar.

6. Place the second set of leaves into the remaining jar. Spray some water containing dissolved nitrate on the leaves, then place the last small beaker with NaOH on top of the leaves. Be careful to avoid direct contact between the leaves and the NaOH solution. Seal the jar.

7. Place all three jars in the location specified by your instructor.

EXPERIMENT 2
Application Lab: Tree Leaves and the Global Carbon Cycle

Pre-Laboratory Assignment   **Due Before Lab Begins**

NAME: _____

Complete these exercises after reading the experiment but before coming to the laboratory to do it.

1. What is the purpose of a control in an experiment? Which sealed jar serves as the control in this experiment? How do you know?

2. Is your mixture in Part III a solution of acids or bases? What color is the phenolphthalein in this mixture?

3. Write balanced equations for the following:

    (a) Hydrochloric acid and sodium hydroxide.

    (b) Hydrochloric acid and sodium carbonate.

    (c) Sodium hydroxide and carbon dioxide.

4. Suppose you compare two of the sealed jars prepared for this experiment last week. Both jar A and jar B contain a beaker with 20.0 mL of 0.908 M NaOH. In addition, jar B contains 3 to 4 dampened leaves. This week you will titrate both solutions with a standardized HCl solution. Circle whichever of the following statements you expect to be true when you do the titrations.

    (a) The NaOH solution in jar A will require more moles of HCl for neutralization than the NaOH solution in jar B.

    (b) Both jars will require the same volume of HCl to reach the end point.

    (c) The NaOH solution in jar B will require more moles of HCl for neutralization than the NaOH solution in jar A.

(d) After adding an appropriate metal ion, no carbonate precipitate will be found in the NaOH solution in jar A.

(e) After adding an appropriate metal ion, no carbonate precipitate will be found in the NaOH solution in jar B.

(f) The solution in jar B contains more dissolved CO_2 than the solution in jar A.

5. In this experiment, you will titrate solution mixtures initially containing carbonate and hydroxide ions. Determine the results in the following two titrations.

(a) The solution containing both ions is titrated with HCl to the phenolphthalein end point. In this case, moles HCl = moles _____.

(b) If the carbonate ion is first precipitated and then the remaining solution is titrated to the phenolphthalein end point with HCl, then in this case, moles HCl = moles _____.

(c) Which of these titrations will you perform on the leaf jars? Why?

6. Explain what you would do if you inadvertently left the stopcock open while filling the buret with the HCl solution and it spilled out over your bench top.

EXPERIMENT 2
Application Lab: Tree Leaves and the Global Carbon Cycle*

Background

Before you analyze your decomposing leaves, you need to determine how you are going to complete the analysis. Your goal in this experiment is to determine the average mass of CO_2 produced per gram of leaf per day. Each sealed jar contains a beaker of sodium hydroxide. Sodium hydroxide solutions absorb CO_2 to produce sodium carbonate and water. CO_2 is present in the atmosphere at all times. The jar without the leaves will reveal the presence of CO_2 upon analysis, and this will serve as a control to determine how much of the CO_2 in the leaf jar is really due to leaf decomposition.

After some time, the solution of sodium hydroxide in the jars will contain sodium carbonate, as a result of absorption of CO_2, along with the unreacted NaOH. Assuming you started out with an excess of sodium hydroxide, you can calculate how much CO_2 reacted by determining how much sodium hydroxide remains unreacted. This will be accomplished by titrating the sodium hydroxide with a solution of hydrochloric acid *after* you precipitate the carbonate from solution. Of course, HCl will eventually react with the solid carbonate: Your technique must allow you to find the end point for the reaction of HCl with NaOH before the HCl begins to react with the solid carbonate.

Part II of this experiment will enable you to determine which reagent is best for the selective precipitation of the carbonate ions in the presence of hydroxide ions in solution. Part III gives you practice with accurate titration of NaOH in the presence of precipitated carbonate ions. Finally, in Part IV, you will analyze your experiment with decomposing leaves.

CAUTION: You will again be working with acids and bases in this experiment. These solutions are corrosive. Wash all small spills (drops) with excess water. Notify your instructor in the event of a large spill (several milliliters). Always wear appropriate safety goggles and wash your hands thoroughly before leaving the lab.

One of the test solutions in Part II may contain a barium compound (it will be labeled if you have it). Many barium compounds are *poisonous*. Handle with care. Dispose of all barium-containing solutions in the designated waste receptacles.

Procedure

Part I: Formation of Groups

Groups of two or three may be used in this experiment. Do *not* begin work on your leaf experiments or your control experiment until you are certain of the best procedure to determine the amount of carbon dioxide. You will have only one chance with each.

Part II: Selective Precipitation of Carbonate Ions (Group Work)

Determine which of the available test solutions reacts to form an insoluble precipitate with sodium carbonate but not with sodium hydroxide. Because you will be using a tissue culture plate for your tests, you only need to use about 10 to 20 drops of each of the solutions per test. Record all observations.

*The method used to analyze for CO_2 from the decay of leaves is described in Anderson, J. P., "Soil Respiration," pp 831–872 in A. L. Page, ed., *Methods of Soil Analysis*, Part 2, 2nd Ed. Soil Science of America and American Society of Agronomy: Madison, WI, 1982.

Results and Conclusions for Part II

Based upon your observations, decide which reagent will selectively precipitate the carbonate ions in the presence of the hydroxide ions (thus leaving the hydroxide ions in solution). Write the net ionic equation for the reaction between the carbonate ion and the reagent you select.

Part III: Titration of $NaOH/Na_2CO_3$ Mixtures (Individual Work)

Prepare a buret for use with HCl.

Obtain a solution containing a mixture of sodium hydroxide and sodium carbonate solutions in unknown amounts. These solution mixtures were made by mixing 1.0 M solutions. Transfer a known quantity of the solution mixture to a 125-mL Erlenmeyer flask. Add two or three drops of phenolphthalein indicator.

Titrate the mixture with the HCl until the indicator color just disappears. The phenolphthalein color fades if left alone because it oxidizes in air. To be sure you have reached the end point, add another drop of phenolphthalein. Record the volume of HCl needed to reach the end point.

Next, transfer another sample of the sodium hydroxide–sodium carbonate mixture to a clean, dry Erlenmeyer flask. Be sure to measure the same volume as used in the first titration. This time, before titrating with HCl, precipitate out the carbonate ion by adding an excess of the metal cation solution you selected in Part II to precipitate the carbonate selectively. To determine how much is an excess, assume the total volume of solution used is only sodium carbonate and calculate the volume of the cation solution needed to completely react with this volume of 1.0 M sodium carbonate. Record your observations.

Add two or three drops of phenolphthalein indicator to your mixture and again titrate with HCl until the color disappears. Verify that you have reached the end point by adding another drop of phenolphthalein.

Calculations for Part III

Calculate the percentage, by volume, of NaOH solution in your unknown mixture.

Results and Conclusions for Part III

Compare the volume of HCl used in both titrations. Explain any differences. If you are not within 5% of the expected value for the percentage of NaOH, repeat your analysis on a different sample of the same unknown. See your instructor for the expected percentage of NaOH of your unknown.

Part IV: Analysis of the Decomposing Leaves Systems (Individual Work)

As the bacteria and fungi on the leaves respired, they produced CO_2. The CO_2 was absorbed into the sodium hydroxide solution, producing sodium carbonate and water. Based upon your work in Parts II and III of this experiment, determine how much sodium hydroxide remains unreacted. Remember to treat the control in the same way you treat the experimental setups.

Calculations

Write equations for each of the reactions that occurred in the analysis of the leaves section of the experiment. You should have three different equations. Calculate the mass of CO_2 produced per gram of dry leaf per day in each of the two jars. Remember, CO_2 is present in the jar even without the decomposing leaves. Your job is to determine how much *more* CO_2 is produced as a result of the leaf decomposition.

Results and Conclusions

Post your group's results for the entire class. Include the type of leaf used and the mass of CO_2 per gram of leaf per day for both plain water and the nitrate-enriched water. Discuss the results in class. Include the class results in your report. Discuss how different conditions may have influenced the results.

The Carbon Cycle

To understand more about the atmospheric levels of carbon dioxide, you could visit *cdiac.esd. ornl.gov/trends/co2/sio-mlo.htm* on the Internet. Click on "Graphics" to view a graph of atmospheric CO_2 concentrations over time. Then return to the previous page and click on "Digital Data" for tabulated values of CO_2 concentrations (in parts per million by volume, or ppmv) each month of every year from 1958–2001. Select several different years and record the atmospheric CO_2 level for each month of those years (record the values directly from the Web site). As a group, you should analyze data from 6 to 8 different years.

Report

Write a summary report to your biologist supervisor explaining your experimental results and your responses to the original questions posed in the scenario. Incorporate your responses to the following questions in your report as well.

1. Using your data and the fact that plants are roughly 44.5% carbon, calculate how many days it would take for all of the carbon in your leaves to be respired if microbes continue to metabolize at the rate you measured. Now think back to a just-fallen leaf outside. Considering how many days you estimate it would take for your leaves to disappear, do you think there are enough days that are moist and warm enough to allow complete metabolism of leaves between fall and spring, say between November 1 and May 15? Explain.

2. Using the data in your table from the Web site, describe the variations in the concentration of CO_2 in the atmosphere throughout the year. Circle the highest and lowest values each year.

3. What two biological processes are responsible for seasonal increases and decreases in atmospheric CO_2 concentration? Atmospheric CO_2 concentrations are a balance between these two processes.

4. Calculate the net change in CO_2 concentration per year. This is the difference between the highest and lowest values in a given year.

5. Atmospheric CO_2 is expressed in ppm by volume on the Web site. Assuming the atmosphere contains 1.8×10^{20} mol of air, using your answer to Question 4, calculate the *net mass* of CO_2 that was added and withdrawn from the atmosphere in a given year.

6. Returning to the data at the Web site, is the difference between the annual high and low atmospheric CO_2 concentrations the same from year to year? (Look at the whole table.) Explain.

7. Is there any evidence that atmospheric carbon is accumulating on land? Use Questions 1 and 6 as evidence.

8. Gasoline and other fossil fuels are primarily hydrocarbons. Therefore, when fossil fuels are burned, CO_2 is produced. Assume that gasoline is composed entirely of octane, C_8H_{18}, and that the burning of gasoline (as in your car engine) can be represented by an equation for the complete combustion of a hydrocarbon. Determine the mass of CO_2 produced by your car in a year. Be sure to record any other assumptions you make in determining your answer. The density of octane is 0.692 g/mL. If you do not operate a motor vehicle, assume you get 19 mpg and drive 11,000 miles per year.

Experiment Group **G**

Buffers and Life: Save a Cardiac Arrest Patient

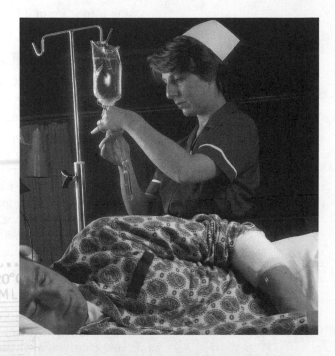

Properly adjusting a patient's fluids can be critical to survival. (James Holmes/Science Photo Library/Photo Researchers.)

Purpose The chemistry of water includes the very important property of acidity and basicity. A solution's acid–base properties can dominate the way that other substances exist and react in water. One measure of the strength of an acid or a base is the amount of hydrogen ion in the solution. This can be followed as a concentration, commonly written $[H^+]$, with units of moles per liter.

The range of $[H^+]$ in solutions is large, from close to 10 to 1×10^{-15} moles per liter. The wide range of possible values for $[H^+]$ means it is often easier to follow a derived function, the pH, of the solution. The pH is a function of the concentration of hydrogen ion, H^+, in the solution, where $pH = -\log[H^+]$. At low pH the concentration of hydrogen ion is high and many substances acquire a hydrogen ion. At high pH many substances will lose a hydrogen ion. This can radically alter the properties of the substance, something that is very obvious in the color of acid–base indicators. It is no surprise, then, that the pH value of a biological system is critical to its operation.

The utility of pH also means that it is important that we know how to stabilize pH. This is done very conveniently with a **buffer.** An acid–base buffer is a mixture that can act to keep the pH of a solution at a nearly constant value. Some buffers can be prepared at very high or very low pH values, using strong acids or bases. But the most common kinds of buffers, the subject of this experiment group, stabilize the pH at intermediate values—between $pH = 2$ and $pH = 11$. These buffers form when both a weak acid and its conjugate weak base are present in solution in significant concentrations.

In this group of labs, you will investigate how buffers work, then apply your understanding to the design of a solution that has certain pH properties. You will also act to correct the pH of a buffer that simulates the blood of a victim of cardiac arrest, because cardiac arrest can cause the pH of the body to change by a dangerous amount.

Schedule of the Labs

Experiment 1: Skill-Building Lab: Stabilization of pH with Buffers

• With a pH meter, follow the effect of added strong acid and strong base on the pH of a solution of a buffer (individual work).

• Characterize the buffer by determining [HA] and [A⁻] (group work).

• Using a mixture of a weak acid and its conjugate weak base, make a buffer that has a particular pH and a particular strength (individual work).

Experiment 2: Application Lab: Cardiac Arrest! The Restoration of Buffers

• Add an unknown amount of strong acid to a buffer, then restore the pH with weak base (group work).

• Assess the buffering ability of solutions for healthy adults (group work).

• Given a solution buffered at an unhealthy pH value, add an appropriate amount of the conjugate base to restore the buffer (individual work).

Scenario

In a hospital emergency room, a patient has just been wheeled in following a cardiac arrest in his home. The paramedics were able to restart his heart after a couple of minutes, and prior to that, CPR was administered so that enough oxygen flowed to his tissues to keep them alive. But his face has the ashen color of someone close to death, and his skin is cold. Though his heart is beating again, he is in danger of death from shock. Immediately, one emergency room nurse starts to prepare an intravenous tube, and another gets a bag of fluid from the supply cabinet. The doctor says, "Yes, let's start him on bicarb" and the nurses begin to administer the fluid, one that helps to restore the victim's blood to its normal state and staves off further damage.

The fluid the ER team has just used to start the stabilization of the patient is the same one that you might use to quiet an upset stomach, or to mildly cleanse a soiled surface. It is sodium hydrogen carbonate (or bicarbonate of soda, also known as bicarb or baking soda). The formula of sodium hydrogen carbonate, $NaHCO_3$, is quite simple, but its role in the body is profound. The hydrogen carbonate ion, HCO_3^-, is a weak base that works with its conjugate weak acid carbonic acid, H_2CO_3, to form a stable pH buffer in the blood. This keeps the blood at a constant pH, something that is essential for life.

The pH of the blood of healthy adults is usually between 7.35 and 7.45. This is maintained by a 1:20 ratio* of H_2CO_3:HCO_3^-. Very small changes in the amount of either substance can radically alter the pH and result in sickness and death. Some of these changes are associated with problems in breathing, such as pneumonia and hyperventilation. In this experiment group, however, we will be more concerned with a patient with metabolic problems.

If the ratio of carbonic acid to hydrogen carbonate becomes too low, then the blood pH will rise and the body will become alkalotic. This can happen, for example, when vomiting occurs and the body uses up carbonic acid to reestablish the stomach acid.

The ratio of carbonic acid to hydrogen carbonate can also be too high, which happens when excess acid is present in the body. The blood pH drops and the body becomes acidotic. Acidosis occurs when, for example, the body is unable to remove carbon dioxide, because the blood stops circulating in cardiac arrest.

When the body starts to become acidotic or alkalotic, there is often enough time for a corrective measure. Most cases of blood pH imbalance are treated by allowing, or by inducing, the body to make the necessary adjustments. This process is known as compensation by the blood. Compensation for pH problems does not correct the ultimate source of the imbalance, but it can prevent other problems from arising.

---

*This ratio depends on normal body pressures and temperatures. While we work at lower temperatures in the laboratory where the ratio is different, the 1:20 ratio is still a helpful guideline.

Certain events cause such dramatic and threatening pH changes that we cannot wait for natural compensation. When a cardiac arrest occurs, the blood stops moving through the body. Immediately, the concentration of carbonic acid starts to rise, lowering the pH. In addition, normal metabolism halts and lactic acid, the product of partial metabolism of carbohydrates, starts to accumulate. This, too, causes the pH to drop. Even after the heart is restarted, the change in the blood's buffer may be so severe that it places the patient in grave danger. Administration of hydrogen carbonate is sometimes the only way to prevent further, perhaps fatal, damage from occurring. Consequently, an understanding of blood pH and buffers is essential to proper methods in the emergency room.

In these experiments, you will progress from studying how buffers work to the point where you are ready to simulate the time-sensitive nature of an emergency room procedure. The material you learn in week 1 will be essential if you are to save your "patient" in week 2.

EXPERIMENT 1
Skill-Building Lab: Stabilization of pH with Buffers

Pre-Laboratory Assignment   **Due Before Lab Begins**

NAME: _____

Complete these exercises after reading the experiment but before coming to the
laboratory to do it.

1. Determine the number of moles of reagent in the following solutions:

   (a) 25.00 mL of 0.10 M acetic acid

   (b) 5.55 mL of 0.092 M NaOH

   (c) 0.50 mL of 0.087 M HCl

2. A buffer solution contains 0.120 M acetic acid and 0.150 M sodium acetate.

   (a) How many moles of acetic acid and of sodium acetate are present in 50.0 mL
   of solution?

   (b) If we add 5.55 mL of 0.092 M NaOH to the solution in part (a), how many
   moles of acetic acid, sodium acetate, and NaOH will be present after the
   reaction is done?

   (c) If we add 0.50 mL of 0.087 M HCl to the solution in part (a), how many
   moles of acetic acid, sodium acetate, and NaOH will be present after the
   reaction is done?

3. Determine the pH you expect to find for the three solutions in Question 2.

4. If you need to prepare 250.0 mL of a pH = 3.60 buffer that has a total buffer concentration of formic acid + formate of 0.030 M, how many moles of each will you need to prepare the solution in Question 1?

5. Discuss how to prepare the solution in Question 4 starting from 0.100 M formic acid and 0.150 M sodium formate.

6. You realize that you have spilled a few drops of sulfuric acid solution on your sleeve, but there is no visible harm. Can you be confident that the fabric is safe? What can you do to protect yourself from such mishaps?

EXPERIMENT 1
Skill-Building Lab: Stabilization of pH with Buffers

Background

Many different technical systems require the control of conditions so that radical changes do not occur when the system is stressed in some way. Here is just a short list, from very different areas of activity:

System	Possible stress	Control mechanism
Temperature of a baking oven	Room-temperature dough placed in the oven.	Large stone "baking tiles" placed in the oven
Suspension bridge	Traffic, some heavy, some light, on the roadway	Suspension cables that can stretch a little
Government	Changes in laws	Constitutional guarantees of rights
International commerce	Currency fluctuations	Central banks able to buy and sell reserves
Body temperature	Hot and cold days	Perspiration, increased metabolism

In all of these situations, there is a control mechanism that can act smoothly to correct the stress, in either one direction or another. The correction may not be complete, but it will lessen, or *buffer*, the system against the stress. However, in each and every case, a large enough stress can overwhelm the control mechanism. In other words, the buffer is not an absolute barrier to change; it is a way of managing stresses of a certain scale.

Chemical systems can also act to buffer against certain kinds of changes in their properties. For example, chemists apply the concept of buffering control to solutions with certain acid–base properties. Mixtures that are able to resist changes in the concentration of hydrogen ion are called acid–base **buffers.** Buffers keep the hydrogen ion concentration, and hence the solution's pH, from varying too much. In some cases, our lives may depend on it.

How do buffers work, how do we characterize a buffer's capacity to resist change, and in what range does a buffer do the "best" job of stabilizing the pH? This skill-building lab will let you observe buffers in action and lead to an understanding of how they are characterized, how they work, and how they fail. In the application lab, you will use this understanding to study a buffer in the human body.

Aqueous Acids and Bases

It is important to review some fundamental ideas about acid–base chemistry before this lab. Your instructor may have other resources for you; use them also if they are available.

The acid and base properties of water come from the presence of two special ions, hydrogen ion, H^+ (this is sometimes written as a complex with water called hydronium ion, H_3O^+), and hydroxide ion, OH^-. Ordinary water contains a small amount of both hydrogen ion and hydroxide ion. The reason for this is that water ionizes to a limited extent, according to the equation.

$$H_2O\ (l) \rightleftarrows H^+\ (aq) + OH^-\ (aq)$$

or

$$2\ H_2O\ (l) \rightleftarrows H_3O^+\ (aq) + OH^-\ (aq)$$

This is an equilibrium reaction. That means that the reaction does not proceed to completion. The concentrations of H^+ and OH^- are related to each other. This is expressed mathematically by an equilibrium expression and an equilibrium constant:

$$K_w = [H^+][OH^-] \qquad K_w = 1.01 \times 10^{-14} \quad (25\ °C)$$

equilibrium expression   equilibrium constant

If we know the concentration of either hydrogen ion or hydroxide ion, then we can calculate the concentration of the other using this expression.

In pure water, the concentrations of hydrogen ion and hydroxide ion are the same, but other chemical substances can affect these. Solutions of acids have more hydrogen ions (and fewer hydroxide ions) than pure water. Solutions of bases have more hydroxide ions (and fewer hydrogen ions) than pure water.

There are many substances that undergo reaction in water to form hydrogen ions. Some acids, called **strong acids,** undergo complete reaction when they dissolve. This means that every molecule of the acid that enters the solution dissociates to give one hydrogen ion. These acids are called strong because they give all the hydrogen ions they can as soon as they dissolve. When one mole of a strong acid is added to water, one mole of hydrogen ions forms.

Some acids do not react completely to give hydrogen ions when they are placed in water. These are called **weak acids.** Some of the weak acid remains unreacted, and some reacts to form less than one mole of hydrogen ion per one mole of acid.

Strong acid: $HA\ (aq) \rightarrow H^+\ (aq) + A^-\ (aq)$
Weak acid: $HA\ (aq) \rightleftarrows H^+\ (aq) + A^-\ (aq)$

Some strong bases react completely in water to give hydroxide ions. For the purposes of this lab, there is only one group of strong bases to consider: hydroxide salts. Many compounds, especially the alkali metal (group I or 1) hydroxides and barium hydroxide, are strong bases. Strong bases give a stoichiometric amount of hydroxide ions when they dissolve in water. This means that *every available hydroxide* in the salt becomes hydroxide ion in the solution, whether the salt contains one, two, or three OH^- ions.

$$NaOH\ (s) \rightarrow Na^+\ (aq) + OH^-\ (aq)$$
$$Ba(OH)_2\ (s) \rightarrow Ba^{2+}\ (aq) + 2\ OH^-\ (aq)$$

There are also weak bases. These react with water to give hydroxide ion, but only partially:

Weak base: $A^-\ (aq) + H_2O\ (l) \rightleftarrows OH^-\ (aq) + HA\ (aq)$

Note that this reaction is *not* the reverse of the reaction of a weak acid with water.

The pH Function

Because $[H^+]$ and $[OH^-]$ can vary by sixteen powers of ten, chemists have found it much easier to express the acid content of a solution by using a logarithmic scale called the pH scale. For a base-ten number system such as the one we use, a logarithm (or log) is defined as the power to which ten must be raised in order to equal the original number. Because most concentrations involve negative exponents, the pH function is defined as a negative logarithm so that most pH values are positive. The pH, then, is the negative log of the hydronium or hydrogen ion concentration:

$$pH = -\log_{10}[H_3O^+] \quad \text{or} \quad pH = -\log_{10}[H^+]$$

Because pH is based on a $\log_{10}$ function, we can easily compare pH values by factors of 10. A solution with a pH of 3.23 has 10 times more hydrogen ion per volume than a solution with a pH of 4.23.

When we know the pH, we sometimes must calculate the hydrogen ion concentration. This is done by "undoing," or performing the inverse of, the log operation. The inverse operation of log is 10^x. Thus, for the function $pH = -\log_{10}[H_3O^+]$, the inverse function becomes

$$[H^+] = 10^{-pH}$$

We can also use a similar function on other variables. For example, when applied to the concentration of hydroxide, we get $pOH = -\log_{10}[OH^-]$. When used with K_w, the value for the equilibrium constant for the ionization of water, we get $pK_w = -\log_{10}K_w$, and for the acid ionization constant K_a we obtain $pK_a = -\log_{10}K_a$.

In addition, we can derive important relationships among the pH, pOH, and pK_w:

$$pOH = -\log[OH^-]$$
$$pK_w = -\log K_w = 13.996 \text{ (at 25 °C)}$$
$$pK_w = -\log[OH^-][H^+]$$
$$= -\log[OH^-] - \log[H^+]$$
$$= pOH + pH$$

The pH scale allows us to quickly label a solution as **acidic** or **basic** relative to neutral water. For pure water at 25 °C, we can determine that the concentration of hydrogen ion is 1.0×10^{-7} M. This gives a pH of 7.00. We say that any solution of water with a pH of 7.00 at 25 °C is *neutral*—even if it is not pure. Water with more hydrogen ions than neutral water, with $[H^+] > 1.0 \times 10^{-7}$, is said to be **acidic**. Acidic solutions at 25 °C have a pH of less than 7. Water with fewer hydrogen ions than neutral water, that is, with $[H^+] < 1.0 \times 10^{-7}$, is said to be **alkaline** or **basic.** Basic solutions at 25 °C have a pH greater than 7.

If we have a solution in which a strong acid controls the amount of hydrogen ions, then it is relatively easy to calculate the value we expect for the pH. This depends simply on the concentration of the acid, because every mole of the acid gives rise to one mole of hydrogen ions. Under those conditions,

$$[H^+] = [HA] \quad \text{Strong acid:}$$
$$pH = -\log[H^+] = -\log[HA]$$

If we have a solution whose pH is controlled by a strong base, then the OH^- concentration is set by the strong base. We can calculate an expected pH from

$$pK_w = pOH + pH$$
$$pH = pK_w - pOH$$

At 25 °C the value of pK_w is 13.996. Therefore, pH = 13.996 − pOH.

Some values for $[H^+]$ and pH are shown on the number line in Figure G-1. Notice something very important. Because of the nature of the pH function, the pH is *smallest*

Figure G-1   The pH scale with representative values for [H⁺]

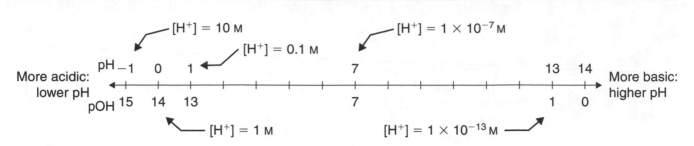

when [H⁺] is greatest. Thus, a solution with a smaller value for pH is more acidic than one with a larger value for pH.

The Equilibrium Expression for a Weak Acid in Water

With **weak acids,** there is incomplete dissociation of the acid. Some of the acid remains intact, and equilibrium is established among the acid, water, the conjugate base, and hydronium ion. As a result, the amount of hydronium ion is *less* than the amount of acid added to the solution.

$$\text{Weak acid: } HA\ (aq) + H_2O\ (l) \rightleftarrows A^-\ (aq) + H_3O^+\ (aq) \quad [H_3O^+] < [HA]_{init}$$

Understanding the relationship among HA, A^-, and H^+ is simple with strong acids, as we have seen. But we need an equilibrium calculation for solutions of weak acids, including their buffers. A weak acid equilibrium requires an equilibrium expression relating the concentrations of the various substances. Following the rules for writing equilibrium expressions, we have

$$HA\ (aq) \rightleftarrows A^-\ (aq) + H^+\ (aq)$$

$$K_a = \frac{[A^-][H^+]}{[HA]}$$

Note the subscript *a* on the *K*. This is a reminder that this expression describes *acid* ionization equilibrium. In other cases, you can use this expression to predict the way that [H⁺] and pH depend on the amount of added weak acid. In this lab group, though, we are interested in how this lets us determine the pH of a buffer.

pH of a Buffer

When we have an acid–base buffer at intermediate pH ranges (between 2 and 11), the pH is controlled by an equilibrium among a weak acid, HA, its conjugate weak base, A^-, and H^+. Here we will discuss how to do the important pH calculations for a buffer.

The equilibrium analysis of buffer concentrations involves the K_a expression, given a known concentration of a weak acid and its conjugate base. If we are certain that neither A^- nor HA reacts very much (less than 5%, a good assumption if $[A^-]$ and [HA] are greater than 0.050 M and the pH is between 2 and 11), then we can use the **Henderson–Hasselbalch** equation to get the pH of a buffer:

$$\text{Buffer: } \quad pH = pK_a + \log\left(\frac{[A^-]}{[HA]}\right)$$

The Henderson–Hasselbalch equation is very helpful in calculating an expected pH for a buffer. If we know the pK_a of a weak acid and the ratio of the conjugate base and the conjugate acid in the buffer, then we can calculate the pH.

There are three types of "stress" we can apply to a buffer: (a) dilution, (b) adding strong acid, and (c) adding strong base. Buffers, though, can resist these stresses. The Henderson–Hasselbalch equation also illustrates how buffers resist changes in pH.

For dilution, the first type of stress, if we have 0.100 L of a buffer that contains 0.025 M HA and 0.050 M A$^-$, then the ratio [A$^-$]/[HA] is 2.0. If the pK_a of HA is 4.00, then the pH will be

$$pH = 4.00 + \log(2.0) = 4.00 + 0.30 = 4.30$$

If we add water to the buffer to increase the volume to 0.120 L, we lower the concentrations to 0.020 M HA and 0.040 M A$^-$, but we *still* have the same [A$^-$]/[HA] ratio of 2.0. The pH should remain at 4.30.

Next, consider what happens when some strong acid is added to the buffer. In that case, some of the weak base is converted to the weak acid:

$$A^- + HCl \rightarrow Cl^- + HA$$

This means that the value for [A$^-$] decreases while the value for [HA] increases. For example, let's say we have another 0.100 L of our buffer that contains 0.025 M HA and 0.050 M A$^-$. If we add 0.020 L of 0.025 M HCl to the 0.100 L of a buffer:

1. We add $0.020 \text{ L} \times 0.025 \text{ M} = 5.0 \times 10^{-4}$ mol of HCl. We also increase the volume to 0.120 L (assuming volumes are additive).
2. The original buffer had $0.100 \text{ L} \times 0.050 \text{ M} = 5.0 \times 10^{-3}$ mol of A$^-$. We will lose 5.0×10^{-4} mol of this in reaction with the HCl, so we are left with 4.5×10^{-3} mol of A$^-$. The concentration of A$^-$ will now be 4.5×10^{-3} mol/0.120 L = 0.038 M.
3. The original buffer had $0.100 \text{ L} \times 0.025 \text{ M} = 2.5 \times 10^{-3}$ mol of HA. We will increase the amount of HA by 5.0×10^{-4} mol through reaction of the HCl and A$^-$, so we wind up with 3.0×10^{-3} mol of HA. The concentration of HA will now be 3.0×10^{-3} mol/0.120 L = 0.025 M.
4. The ratio [A$^-$]/[HA] is now 0.038/0.025 = 1.5, so the pH is calculated to be

$$pH = 4.00 + \log(1.5) = 4.00 + 0.18 = 4.18$$

Despite the addition of 20. mL of strong acid, the pH changes very little!

A similar control is exerted when strong base—for example, NaOH—is added to a buffer. We can use the same reasoning we just used for the strong acid example. Briefly, if we have 0.100 L of buffer with 0.025 M HA and 0.050 M A$^-$ and add 0.020 L of 0.025 M NaOH, then [HA] will decrease to 0.017 M and [A$^-$] will increase to 0.046 M. The ratio [A$^-$]/[HA] is now 0.046/0.017 = 2.71, and the pH should be 4.43.

Analysis of Buffer Capacity: What We Look For

There are three important questions we must ask about a buffer solution:

1. What is the pH and, therefore, [A$^-$]/[HA]?
2. What is the actual concentration of [HA]?
3. What is the actual concentration of [A$^-$]?

To answer the first question, a simple pH measurement and the Henderson–Hasselbalch equation are enough. To answer the other questions, we need to use acid–base titration to determine the buffer's capacity.

At a certain point in the addition of acid or base to the buffer solution, the contents of the buffer will be overwhelmed. The point where that occurs in this experiment can be used to determine the concentration of the components of the buffer.

When the addition of a strong acid or a base overwhelms the buffer, the pH changes rapidly with even a small amount of acidic or basic solution. This is known as the *end point* of the reaction (even though you should collect data beyond that point!). The end point should, if you are careful, be close to the *equivalence point* of the reaction—the point when the amount of added acid is exactly equal to the amount of the base in the original solution, or vice versa. At the equivalence point, assuming that the acid and the base both donate or accept just one hydrogen ion, the following is true:

Equivalence point when adding strong acid to the buffer:	**Equivalence point when adding strong base to the buffer:**
Moles base in buffer = moles added acid	Moles acid in buffer = moles added base
$c_{base}V_{base} = c_{acid}V_{acid}$	$c_{acid}V_{acid} = c_{base}V_{base}$
$c_bV_b = c_aV_a$	$c_aV_a = c_bV_b$

These equations are very helpful, but they can be used in the wrong situation. In particular, they apply *only at the equivalence point* and *only for titrations where the stoichiometry of acid to base is 1:1.*

For both of these equations, you should know the volume (V, in L), the concentration (c, in mol/L) of the added solution, and the volume of the buffer. It is then possible to solve for the concentration of the acid or the base in the buffer.

Methods of Buffer Design and Preparation

There are two methods of preparing buffers. One involves adding a strong acid to a weak base or a strong base to a weak acid. You will not use that method today. Instead, you will simply mix the weak acid and base directly. To do this, you mix appropriate amounts of a weak acid and a weak base, then dilute to the correct final volume.

CAUTION: Handling solutions of acids and bases requires special precautions. Though they do not stain clothing, a few drops of either substance can harm fabric, something that is only apparent after laundering. The solutions of NaOH that you are handling pose a particular risk to the eyes.

Many different solutions will be in use today. If one disposes of an acid or a base in the wrong manner, then mixing of acids and bases may occur in a very short time period. This has the potential to release a large amount of heat. Therefore, label solutions carefully and dispose of them in the proper container. If necessary, neutralize both acids and bases with sodium hydrogen carbonate or a similar reagent before putting them into a common waste receptacle.

Procedure

This experiment should be done by students individually. If needed, students may share burets, but each student must record a separate set of data for each part, to ensure familiarity with the pH meter and the burets.

Each student will record two tables of data in this laboratory:

1. Addition of strong acid to a buffer solution.
2. Addition of strong base to a buffer solution.

It is essential that the notebook show both the values of any buret reading and the pH reading. All buret readings should be made to ±0.02 mL. This requires reading between the lines on the buret, a good measurement practice.

When adding strong base or acid to a buffer solution, we suggest the headings in Table G-1. Don't forget to include a table entry for the solution *before* addition of the acid or the base.

TABLE G-1

Initial Data:
V (buffer): _____
Concentration of strong weak acid: _____
Concentration of strong weak base: _____

Buret reading	pH	Color

Part I: Calibration of pH Meter (Group Work)

Obtain a pH meter from your instructor. Be certain to handle the pH probe carefully, for its glass tip can be very fragile. Also obtain approximately 15 mL of a standardization solution in a clean, dry beaker. This solution is prepared to have a known pH value; you will adjust your pH meter to the value of the standard.

If necessary, plug in the meter and set it to measure pH. Turn on the meter and allow it to warm up for at least 3 min. Gently rinse the tip of the probe with a stream of deionized water, then place the probe in the standardization solution. You should choose a standardization solution within 2.0 pH units of the range you will be studying. After allowing the probe and the solution to reach the same temperature, adjust the meter until it reads the correct value for the pH.

Part II: Addition of Strong Acid and Base to Standard Buffer (Group Work)

In this part of the experiment, you will record the pH changes that occur when strong acid and strong base are added to a buffer solution containing sodium propionate and propionic acid.

Place 25.00 mL of the buffer and five drops of bromocresol green in a beaker, with a stirring bar. Begin stirring, then place the pH probe into the solution. Add the strong acid in approximately 0.50-mL increments. It is not necessary to add *exactly* that volume in each case, but you should aim for an increment around that value. In any event, record the actual buret reading (to 0.02 mL), pH, and color of the solution at each point. Use the format of Table G-1. Continue until the pH drops below 2.00.

Repeat this same procedure with a fresh sample of buffer and a clean beaker; use strong base this time and continue adding increments until the pH rises above 11.00. If there is time, repeat each titration.

Part III: Designing and Preparing a Buffer (Individual Work)

You will be assigned a buffering agent (ammonium–ammonia, lactic acid–lactate, acetic acid–acetate), a total buffer concentration, and a target pH. Your partners will have different buffers to examine, but there will be similarities. Work together to design your buffers, making note of when your calculations are the same and when they are different.

You should prepare 250.0 mL of buffer at the target pH by mixing the weak acid and the weak base. Dilute to the correct final volume. Check the pH. The following example outlines this procedure.

Example: Prepare 0.250 L of a pH = 4.00 solution of formic acid with a total concentration of 0.0400 M. Formic acid has a pK_a of 3.74. Assume we have stock solutions with 0.100 M formic acid and 0.100 M sodium formate. Although it is conceptually simpler to say, "Mix the correct amount of the acid and the base," we have to do much more in the initial calculation.

1. *Determine the correct [HCOOH]/[HCOO⁻] ratio.* This requires us to rearrange the Henderson–Hasselbalch equation:

$$pH = pK_a + \log\left(\frac{[HCOO^-]}{[HCOOH]}\right)$$

$$pH - pK_a = \log\left(\frac{[HCOO^-]}{[HCOOH]}\right)$$

$$\frac{[HCOO^-]}{[HCOOH]} = 10^{(pH-pK_a)} = 10^{(4.00-3.74)} = 10^{+0.26} = 1.82$$

This means that the concentration of formate should be 1.82 times that of the concentration of formic acid: $[HCOO^-] = 1.82\ [HCOOH]$

2. *Determine the actual concentrations of formate and formic acid needed to make the 0.0400 M buffer.* We know we must have the two concentrations add to the total buffer strength of 0.0400 M. Therefore,

$$[\text{buffer}] = [HCOOH] + [HCOO^-]$$

Substituting and solving for [HCOOH]:

$$0.0400\ \text{M} = [HCOOH] + 1.82[HCOOH]$$
$$0.0400\ \text{M} = 2.82[HCOOH]$$
$$0.0142\ \text{M} = [HCOOH]$$

Now that we know [HCOOH], we can solve for [HCOO⁻]:

$$[HCOO^-] = 1.82[HCOOH] = 0.0258\ \text{M}$$

3. *Mix the correct mole and volume amounts of formic acid and a formate salt to give the desired concentration.* We have determined that if we made 1 L of solution we would need 0.0142 mol of formic acid and 0.0258 mol of formate. Since we are going to make 0.250 L of solution, we need to mix 0.00355 mol of formic acid and 0.00645 mol of formate. We will get these from the stock solutions of 0.100 M in each component:

Volume of formic acid: $0.00355\ \text{mol HCOOH} \times \dfrac{1\ \text{L}}{0.100\ \text{mol}} = 0.0355\ \text{L formic acid}$

Volume of formate acid: $0.00645\ \text{mol HCOO}^- \times \dfrac{1\ \text{L}}{0.100\ \text{mol}} = 0.0645\ \text{L formate}$

Thus, we mix 35.5 mL of formic acid solution and 64.5 mL of sodium formate solution in a 250.00-mL volumetric flask, then add water with mixing to give a total solution volume of 250.00 mL.

Report

All sets of data should be graphed with volume of HCl or NaOH on the x-axis and the pH on the y-axis.

Determining Buffer Characteristics

Your report should conclude with answers to the three questions we posed earlier about buffers: (1) initial pH and $[A^-]/[HA]$, (2) concentration of HA, and (3) concentration of A^-. The answers to (2) and (3) will help you determine the capacity of the buffer. Depending on the amount of the weak acid and the weak base present, this buffering capacity will eventually be overwhelmed at the equivalence point. This should be apparent from the experimental data curve, in the form of a much larger change in the pH.

The volume of strong acid or strong base added to reach equivalence can be used to determine the number of moles of HA and A^- in the buffer. These, along with the volume of the buffer, give $[HA]$ and $[A^-]$. Obtain the actual HA and A^- concentrations from your instructor. Calculate the percentage error in your results and discuss the reasons for any differences in the concentrations. If time permits, you may ask to repeat the titration and recalculate your results.

A final question to answer is the total concentration of the buffer. This refers to the concentrations of $[A^-]$ and $[HA]$ in the initial buffer.

$$\text{Buffer concentration} = [HA] + [A^-]$$

EXPERIMENT 2
Application Lab: Cardiac Arrest! The Restoration of Buffers

Pre-Laboratory Assignment   **Due Before Lab Begins**

NAME: _____

Complete these exercises after reading the experiment but before coming to the laboratory to do it.

1. Use the Henderson–Hasselbalch equation to determine the ratio of acid to base in a formic acid–formate buffer with pH = 3.00.

2. Assume that the human blood buffer includes, at any one point, 0.00080 M carbonic acid and 0.0016 M hydrogen carbonate. Determine the number of moles of each component present in a person with 7.00 L of blood. What is the ratio of carbonic acid to hydrogen carbonate in this case?

3. Metabolic acidosis results in the addition of excess acid to blood. How many moles of acid must be added to the blood in Question 2 to bring the carbonic acid/hydrogen carbonate ratio to the hazardous level of 1:10?

4. Mixing acids and bases rapidly can be very dangerous. Before pouring any acid or base in a disposal unit, what can be done to neutralize the solution?

EXPERIMENT 2
Application Lab: Cardiac Arrest! The Restoration of Buffers

Background

As we outlined at the beginning of this experiment group, the buffering of the blood is important in maintaining health. If the pH of the blood drops below 6.9 or rises above 7.8, then death is likely. But even within the "safe" range, the optimum pH for many bodily processes is quite narrow. Therefore, slight disturbances to the pH can significantly impair normal processes. And, of course, these may be the same processes that are used to recover after a critical incident such as cardiac arrest.

When the heart stops, two important problems arise that threaten the blood pH. First, metabolic acidosis sets in. Normal metabolism requires oxygen, which converts a molecule such as glucose all the way to carbon dioxide:

$$C_6H_{12}O_6 + 6\,O_2 \rightarrow 6\,CO_2 + 6\,H_2O$$

When the blood stops providing oxygen to the tissues, however, the tissues can only carry out metabolism partway, through the breakdown of glucose into lactic acid:

$$C_6H_{12}O_6 \rightarrow 2\,H_3C\!-\!CH(OH)COOH$$

The lactic acid generated in this way is the same lactic acid that can build up in the muscles during anaerobic exercise. This causes a familiar aching and fatigue in the muscles, which of course encourages us to rest. However, during cardiac arrest this reaction occurs in *all* the tissues of the body, and they are not equipped to handle a lactic acid buildup. So they dump it, right into the blood stream. The lactic acid then reacts with the hydrogen carbonate ion. This converts some of the blood's HCO_3^- to H_2CO_3 and alters the H_2CO_3:HCO_3^- ratio.

$$C_3H_6O_3 + HCO_3^- \rightarrow H_2CO_3 + C_3H_5O_3^-$$

A second problem is that there is some carbon dioxide in the tissues that is formed by the last bits of available oxygen. This, too, is dumped into the blood, but it cannot be exhaled because the heart has stopped and blood no longer moves to the lungs. Even after the heart is restarted, there can be quite a delay before the excess CO_2 is removed.

Thus, it is a common protocol to quickly administer hydrogen carbonate to cardiac arrest patients. But emergency room personnel don't have the time to "titrate" the patient. They must administer a reasonable amount right away, then check to see if the treatment has worked.

The carbonic acid–hydrogen carbonate buffer is difficult to study in the general chemistry laboratory because it is too easy to lose carbon dioxide to the room atmosphere. Therefore, in this lab you will work with a buffer that behaves similarly to the carbonic acid buffer: the buffer formed between the acid dihydrogen phosphate, $H_2PO_4^-$, and the conjugate base hydrogen phosphate, HPO_4^{2-}.

In this experiment, you will eventually be confronted by a solution of a simulated blood buffer that has been disturbed by cardiac arrest. To be ready for this challenge and to increase your understanding of buffers, you will do the following tasks:

- Show that you can restore a buffer to within an expected pH range after it is "challenged" by the addition of a strong acid.

- Characterize the buffer designed for the blood by determining its pH and the concentration of the weak acid and weak base that are present.
- Show that you can restore the blood buffer when it has been challenged by the results of a cardiac arrest.

Determining What Has Happened When a Buffer Is Disturbed

This experiment focuses on restoring buffers after a disturbance has made the buffer too acidic. Using information about the undisturbed buffer, it is possible to find out exactly what happened. Counteracting this is the basis of a plan to restore the buffer.

Let's say we have 100.0 mL of a buffer with a total buffer concentration of 0.075 M with a pK_a of 4.00 (you may recognize this as the buffer we used to illustrate our calculations of buffer behavior on page G-11). The undisturbed buffer had a pH of 4.30 with [HA] = 0.025 M and $[A^-]$ = 0.050 M. This meant that the ratio of $[A^-]/[HA]$ = 2.00. But now we find that the pH is 4.15 instead, because someone has added some strong acid to the solution, converting some A^- to HA. We need to determine the new concentrations of HA and A^- and then restore their ratio to the correct value.

In this situation, one thing has not changed: the total concentration of the buffer. Therefore

$$[HA] + [A^-] = 0.075 \text{ M}$$

We can also calculate the ratio of $[HA]/[A^-]$ by rearranging the Henderson–Hasselbalch equation:

$$pH = pK_a + \log\left(\frac{[A^-]}{[HA]}\right)$$

$$pH - pK_a = \log\left(\frac{[A^-]}{[HA]}\right)$$

$$10^{pH-pK_a} = \left(\frac{[A^-]}{[HA]}\right)$$

$$10^{pK_a-pH} = \left(\frac{[HA]}{[A^-]}\right)$$

We now have two equations in two unknowns. Substitution will allow us to calculate the actual concentrations of HA and A^-.

To restore this buffer to the undisturbed pH, we need to add enough A^- to bring the ratio back to the correct 0.50 value. We will need to calculate how the actual $[A^-]$ compares with the value in the undisturbed buffer. The difference is the amount by which $[A^-]$ needs to change.

CAUTION: In this lab, you are working with solutions that are near the normal pH of blood—which is close to pH = 7. Although this pH is considered a "neutral" solution in chemistry (at 25 °C), your solutions may become much more acidic or basic. Use a small piece of pH paper to be sure that a solution is neutral before disposing of it.

Procedure

Part I: Formation of Groups

This laboratory will be carried out by individuals disturbing buffers made from weak acids and weak bases. Each student will give some of his or her solution to another student, who will add a small amount of HCl to lower the pH by an arbitrary amount. The first student will then attempt to add enough weak base to restore the buffer. Finally, the two partners will characterize "normal" blood and restore blood after a cardiac arrest.

Part II: Disturbing and Restoring the Buffer (Group Work)

Your partner will take 100. mL of a standard buffer and add enough strong acid to lower the pH by between 0.20 and 0.40 pH units. He or she should record the amount of HCl added to do this. The buffer will be returned to you and you should determine what has happened in the disturbance. This will require a pH measurement and the Henderson–Hasselbalch equation to determine the new ratio of $[A^-]/[HA]$. You can then calculate how many moles of A^- have been converted to HA during the disturbance. Compare your calculation with the amount of HCl your partner actually used. Your values should agree to within 10%. If not, repeat the disturbance and calculation with another 100. mL of buffer.

The amount of HCl added has disturbed your original ratio of $[A^-]/[HA]$. You need to add enough A^- to restore the ratio. Calculate how much A^- solution is needed to do this, and add that amount to the disturbed buffer. Your final pH should be within 0.10 pH unit of the original, undisturbed value.

Part III: Characterization of the "Blood" Buffer (Group Work)

Before you use your knowledge of buffers to save a patient, you need to know what is in the blood's buffer. Each pair of students should obtain some "normal" blood buffer. Characterize it completely, as you did in the skill-building lab.

Part IV: The Treatment Protocol (Group Work)

Now that you know what normal blood contains, you can try to restore some of it, just as you did with your buffer in Part III of this experiment. Now, however, your work mode is saving a cardiac arrest victim. Consider the following critical care question:

> A patient has a body mass between 75 and 225 lb. The patient's blood pH, due to the disruption of the blood buffer, is now between 7.00 and 7.25. How many mL of bicarbonate must be added to bring the patient's blood pH to between 7.35 and 7.45?

We will not be able to use real patients or the volume of blood in real people, but we will simulate the problem with the following substitution:

> A patient has a blood volume between 75 and 225 mL. The patient's blood pH, due to the disruption of a $H_2PO_4^-$—HPO_4^{2-} buffer, is now between 7.00 and 7.25. How many mL of 0.050 M HPO_4^{2-} must be added to bring the patient's blood pH to between 7.35 and 7.45?

To answer these questions, your group will receive a sample of "blood" with a pH that is too low for a healthy patient. You will determine the amount of HPO_4^{2-} necessary to restore your patient's blood pH. Now, however, you will get only *one* chance to do this (just as might occur in a real emergency!).

Part V: Save Your Patient! (Individual Work)

Go to your instructor to get a slip, which will direct you to one of the patient solutions in the lab. Take only the indicated volume from the stock bottle. Bring it to your work area, measure the pH, then calculate how much HPO_4^- to add in order to restore the pH to 7.35–7.45. Your instructor will write down the time at which you got the slip. You will have 30 minutes to save your patient.

When your calculations are done, prepare the necessary volume of HPO_4^-. Ask your instructor to watch as you add *all at once* the amount of HPO_4^{2-} you think you need. Mix well and ask the instructor to measure the pH with a pH meter. If your pH is in the target range, congratulations! If not, remember, this is only a lab!

Experiment Group H

Measurement of Chemical Reaction Rates: Clean Up Waste Water

Determining the contents of waste water is important to environmental safety.
(Alan D. Carey/Photo Researchers.)

Purpose Determining the rate of a reaction and discovering inexpensive ways to increase that rate are important in a large number of chemical systems, from the simplest combustion process to the most complex biological transformation. Therefore, studying reaction rates, or reaction kinetics, is critical in all areas of chemistry and their application. This group of experiments will introduce you to the kinds of problems that chemical engineers face in their work in studying and using chemical reactions.

In the first lab, the method of spectrophotometry is used to characterize the formation of a colored product of a reaction. In the second lab, you will monitor a reaction that consumes a colored substance, or dye. This will simulate an important method of removing contaminants from water supplies.

Schedule of the Labs EXPERIMENT 1: Skill-Building Lab: Hydrolysis of an Ester
- Determine the reaction rate for the hydrolysis of *para*-nitrophenyl acetate (group work).
- Examine the effect of added nitrogen-containing compounds on the rate (group work).

EXPERIMENT 2: Application Lab: Decomposition of an Aqueous Dye
- Design a procedure to follow the decomposition of a dye in a mixture with iron filings (group work).
- Evaluate the rate of reaction for different amounts and types of iron (individual work).

Scenario Here is a dialogue that shows how an engineer, working with chemistry, can encounter the problem of reaction rates in industry.

"Oh, boy," your boss at ACME Chemical Company says, "the Feds are really starting to crack down on our waste releases. If we don't find a way to treat our exit streams, we're going to get fined to the annual tune of $20,000 per ton of waste released."

"Okay, here you go," your boss says. "Our archrival, Megachemco, has just come out with a great new compound for decomposing dye in a waste stream. Speeds up the rate a million times over anything we've seen. Their reactor costs are now smaller than my daddy's income tax payments."

You respond, "What do we know about their reactant?"

"Their patent claims that 20 grams of the stuff can decompose 95% of a 0.01 M solution of dye in a hundred-gallon reactor in twelve minutes. Our marketing people have placed calls to Megachemco; they tell us that the reagent costs a fortune at $5,000 per kilogram."

"Then I've got an idea," you immediately brainstorm. "We don't necessarily need a reactant that's more active than Mega's. If we can find one that's, say, half as active but costs 100 times less, we can add a lot more reactant and get higher activity for the same amount of money."

"Hot dog," shouts your boss, pounding his fist on the desk, "I knew you'd think of something. You know, we have lots of iron shavings and powder stockpiled around here. Why don't we try those first? Find out the particulars: For the same price per batch, how long will it take these different kinds of iron to decompose 95% of the dye under Mega's conditions? By what percentage can we cut down their processing time?"

This is a real problem, you realize, because the Acme Chemical company, and more specifically, the process stream you are in charge of, releases 3.5 tons of dye a year (on a daily basis, that's 19 lb/day). It's diluted to very low levels before release, of course, well within current regulatory limits, but apparently the regulations are being tightened.

"My raise, my promotion!" you think to yourself. "If I can't make Acme money with my process stream, I'm the one who's going to get in trouble."

Yet you also know that the present state of ecological awareness is not a bad thing for society. There are current initiatives in "green syntheses," which use environmentally benign substances in reactions that generate little or no waste.

"Let's get down to specifics," your boss continues. "Remember the ten-gallon holding tank we're currently bypassing? Let's make it into a batch tank reactor. Shoot the waste into that and react it with something to decompose it. Then it will be safer to dispose of, and we'll stay way ahead of the regulations."

You reply, "Well, we can't bleach it, because that may chlorinate the aromatics in the waste, which would make it worse. You're right about the iron, too."

"You new college grads are great at seeing the big picture," your beaming boss admits. "Find out how many times a day we'll need to run the reactor, what the least amount of time (and cost) per run is, and the total annual net cost. Oh, by the way, I'll give you 10% of the difference between the cost of the fine and the cost of the clean-up as your next year's raise."

The chemical industry has provided many benefits to society, typically through the synthesis of new compounds with beneficial properties. But the "other side" of chemistry is the production of waste that must be disposed of in an efficient and complete manner.

Oxidation–reduction reactions are an important class of chemical reactions. One of the best ways to carry out an oxidation reaction is with a chlorine derivative, such as chlorine bleach. However, this has the potential to produce a large number of waste products containing organochlorine groups, which are implicated in a number of environmental problems, so green synthesis seeks to replace chlorine oxidants with nonchlorine oxidants or with reducing agents.

A final consideration for this scenario is the design of a system that will work on a waste effluent that will be discharged from a synthetic reaction. Such effluents are not collected: The unwanted materials must be removed as they flow out of a reaction system, but before they reenter the environment.

Chemical engineers are constantly facing challenges such as these. They must have a mastery of basic chemistry, and they must be able to design processes that are efficient and reliable. One component of the design is that the reactions must be continuous. You will gain experience with this in this group of experiments. While the reactions are done in "batch mode" (that is, one at a time), you will get data about the progress of the reaction by continuously checking the change in one parameter of the system while not disturbing the reaction itself.

EXPERIMENT 1
Skill-Building Lab: Hydrolysis of an Ester

Pre-Laboratory Assignment   **Due Before Lab Begins**

NAME: _____

Complete these exercises after reading the experiment but before coming to the laboratory to do it.

1. The reaction you are studying generates acetic acid as one of the products. This could affect the pH of the solution. What precaution is taken against this?

2. Sketch graphs for the dependence of the reaction absorbance A with time for a reaction that is (a) zero order in PNA and (b) first order in PNA.

3. A buffer is prepared from 0.30 M phosphate stock solution, and 25.00 mL of this is diluted to a final volume of 100.00 mL. What is the concentration of the phosphate buffer in this solution?

4. The buffer for this reaction contains dihydrogen phosphate and hydrogen phosphate. Dihydrogen phosphate has a K_a of 6.23×10^{-8}. At pH = 7.00, what is the ratio of [dihydrogen phosphate] to [hydrogen phosphate] in this solution?

5. What should be done to dispose of the reaction solutions in this experiment?

EXPERIMENT 1
Skill-Building Lab: Hydrolysis of an Ester

Background

Hydrolysis Reactions

The reaction of a compound with water is very important in chemistry, especially in biological systems. In many cases, such a reaction results in a *splitting*, or *lysis*, of the compound into two parts. Such "splitting-by-water" reactions are called *hydrolysis* reactions.*

Many different molecules contain a group of atoms called an ester, and these can be hydrolyzed by water to make a —COOH group (also called a carboxylic acid) and an HO— group (which, when the HO is attached to a C atom, are usually called alcohols).

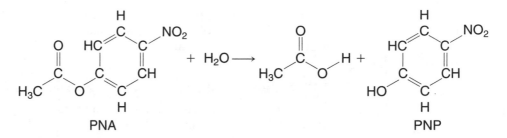

In the diagram above, R and R' can refer to any group, such as H, CH_3, C_6H_5, etc.

This reaction is spontaneous for almost all esters. However, it can be very slow under typical conditions of temperature and pressure. It is much faster if there is a significant amount of a base in the solution, whether hydroxide ion or another base.

In this week's experiment, you will study the rate of this reaction using an ester that produces an alcohol, *para*-nitrophenol (PNP). The ester is called *para*-nitrophenyl acetate, or PNA.

PNA   PNP

Under the conditions studied here, both the acetic acid and the *para*-nitrophenol lose hydrogen ions to make anions in steps that are fast relative to the hydrolysis of the ester itself. The PNP anion has a characteristic yellow color due to its absorption of light at 405 nm.

> **CAUTION:** Neither *para*-nitrophenol nor *para*-nitrophenyl acetate is listed as a carcinogen by the National Institutes of Health, but they have been found to cause chromosomal abnormalities. Therefore, they should be handled with extreme care. Proper gloves, careful washing, and disposal in designated waste containers are essential.
>
> The phosphate buffers and catalyst solutions are potentially irritating. Thoroughly wash any skin areas that come into contact with these and clean up any spills immediately.

---

*There are other kinds of hydrolysis reactions, including the reaction of acids and bases with water.

Procedure

Part I: Formation of Groups

You will work in groups of two, with each pair of students preparing and studying one phosphate concentration and one catalyst. Later on, the group results will be pooled to share data about two different phosphate concentrations and several different catalysts.

Each group will have several different solutions in use at any one time. Carefully label all beakers and flasks.

Because of the dramatic effects of catalysis, be certain to wash all glassware carefully before using. Pipets should be used for one solution at a time; do not use the same pipet for different solutions without careful cleaning and pre-rinsing.

Part II: Preparation of Phosphate Buffer Solution (Group Work)

Each pair of students will examine the rate of the reaction at a different concentration of a phosphate buffer. The buffer contains both hydrogen phosphate and dihydrogen phosphate ions, with the sum of their concentrations equal to the amount of phosphate in the buffer. Thus, when we say that a solution is 0.10 M phosphate buffer, we do not mean that the concentration of PO_4^{3-} is 0.10 M. We mean that the concentration of all phosphate-related substances and ions (phosphoric acid, dihydrogen phosphate, hydrogen phosphate, and phosphate ion itself) add to 0.10 M. The actual concentration of the components depends on the pH of the solution.

All experiments by a pair of students will be done with the same buffer, so all solutions must be prepared from the same buffer solution.

Obtain 25.00 mL of phosphate buffer (either 0.20 M or 0.40 M) in a 50- to 100-mL beaker. This will have a pH of about 6.50, and you have to adjust the pH to your assigned value. Place a *calibrated* pH probe in the solution, then add 2 M KOH solution *dropwise* to bring the pH to your target value. When you are done, transfer all of your solution to a clean 100.00-mL volumetric flask that has been rinsed twice with deionized water. Rinse the beaker with a few milliliters of deionized water and add the rinsing to the volumetric flask. Finally, add deionized water to bring the total solution volume to 100.00 mL.

Part III: Determination of the Rate of Hydrolysis in Buffer (Group Work)

In this experiment, you will measure the appearance of *para*-nitrophenol (PNP) in a solution. Standardize the spectrophotometer with a blank containing only the phosphate buffer, using the reference wavelength of 405 nm. You will then study the rate of hydrolysis in the buffer itself. This slow reaction will show the importance of the catalyst.

Obtain about 20 mL of PNA solution in a clean, dry beaker. Note the concentration. This solution is stable as prepared, but it begins to react when it is put in a solution above pH = 6.5. Therefore, do not add PNA to any solution until you are ready to begin the measurement of the appearance of PNP.

You start the reaction by mixing a certain amount of buffer with a fixed amount of the PNA solution. The exact values will depend on the volume of the spectrophotometric cell you are using. If you are using cuvettes, then the total volume is about 2.80 mL. Use a Mohr pipet to add 2.30 mL of the buffer to a cuvette. When you are ready to begin the experiment, add 0.50 mL of the PNA.

To mix the solutions and start the reaction, cover the cuvette securely and invert it to mix the solutions well, or use a Pasteur pipet to quickly withdraw and reinject the solution in the cuvette several times.

Try to take the first absorbance measurement at 30 s after mixing, and then take additional measurements every 5 min for 30 min. When you are done, you will need to take one more absorbance measurement, when absolutely all the PNA has reacted. To do this, add *one drop* of the "supercatalyst" solution your instructor has. Mix well, then wait 15 min for the hydrolysis to go to completion.

While this reaction is proceeding, one of the students in the pair can prepare the two solutions of the catalyst candidate needed for Part IV of the experiment.

Part IV: Determination of the Rate of Hydrolysis with a Catalyst (Group Work)

Your instructor will assign a catalyst to each group. Obtain 10 mL of the catalyst solution in a clean, dry beaker. Note the concentration of the catalyst.

Prepare two solutions of the catalyst in the buffer. For the first, transfer 2.50 mL of the catalyst and 2.50 mL of deionized water into a 25.00-mL volumetric flask, then dilute to the mark with your phosphate buffer. For the second, use 5.00 mL of the catalyst in a second volumetric flask, then dilute with your buffer.

Prepare hydrolysis solutions as before: Use the *same* amount of buffer + catalyst and PNA solution as in Part III.

If a pair of students plans well, then all three samples—the phosphate buffer only and the first and second catalyst solutions—can be studied at the same time by careful switching of the solutions in the spectrophotometer. All solutions should be measured every 5 min for 30 min.

Report

Your data need to be analyzed for the increase in the concentration of a product PNP over time. However, you will not have to determine the absolute concentration, which would require a multistep calculation. Instead, you can determine the rate of the reaction by looking at the absorbance values only.

If we have a reaction in which only one species, a product, absorbs light, then we can follow the increase in the absorbance to determine the reaction rate. This assumes that the species obeys Beer's law under the conditions of the experiment. In that case, $A = \varepsilon c l$, where ε is a coefficient for the amount of light absorbed, c is the concentration of the substance, and l is the length of the cell. In this experiment, we do not determine c itself. Instead, we follow the relative amount of the product by comparing A at different times.

To get the reaction rate, we analyze the reaction using first-order kinetics in the PNA. This means the rate $= k[\text{PNA}]$.

Using the appropriate integrated rate law, we get a dependence of the concentration of PNA on the time: $[\text{PNA}] = [\text{PNA}]_0 e^{-kt}$. At time $t = 0$, $[\text{PNA}] = [\text{PNA}]_0$ and at time $t = \infty$, $[\text{PNA}] = 0$.

In this reaction, PNP is directly produced from PNA in a 1:1 ratio. Therefore, the final concentration of PNP is equal to the initial concentration of PNA. At any time t between the start and the finish,

$$[\text{PNP}] = [\text{PNA}]_0 - [\text{PNA}]$$
$$= [\text{PNA}]_0 - [\text{PNA}]_0 e^{-kt}$$
$$= [\text{PNA}]_0 \, (1 - e^{-kt})$$
$$= [\text{PNP}]_{\text{infinity}} \, (1 - e^{-kt})$$

Assuming that Beer's law is obeyed, $[\text{PNP}] = A/\varepsilon l$. Therefore,

$$A_t/\varepsilon l = A_{\text{infinity}}/\varepsilon l (1 - e^{-kt})$$
$$A_t = A_{\text{infinity}}(1 - e^{-kt})$$
$$A_t/A_{\text{infinity}} = (1 - e^{-kt})$$
$$(A_t/A_{\text{infinity}}) - 1 = -e^{-kt}$$
$$1 - (A_t/A_{\text{infinity}}) = e^{-kt}$$
$$\ln[1 - (A_t/A_{\text{infinity}})] = -kt$$

If we plot $\ln[1 - (A_t/A_{infinity})]$ versus t, we will get a line with a slope equal to $-k$.

Using your data, prepare a table that has time for the first column, A in the second column, and $\ln[1 - (A_t/A_{infinity})]$ in the third column. Use these values to prepare graphs of A_t versus t and $\ln(1 - A_t/A_{infinity})$ versus t. Determine the slope m for this line and calculate k. To assess the effect of the catalyst, compare the graphs for different catalyst concentrations. The *relative* slopes of these graphs will provide the order of the reaction in the catalyst.

EXPERIMENT 2
Application Lab: Decomposition of an Aqueous Dye

Pre-Laboratory Assignment   **Due Before Lab Begins**

NAME: _____

Complete these exercises after reading the experiment but before coming to the laboratory to do it.

1. In the skill-building lab, you followed the increase in absorbance at a wavelength of 405 nm. In this experiment, different students have different dyes and therefore different wavelengths to follow. Explain how you will determine the best wavelength for your dye.

2. Imagine you have a dye with an absorbance of 0.542 at a particular wavelength when the dye is at a concentration of 4.2×10^{-3} M. After 10.0 min, the absorbance is 0.123. Assuming Beer's law applies, what is the concentration of the dye after 10 min?

3. Can you determine the reaction order using the data in Question 2? What else must you do to determine the reaction order?

4. What safety concern must be addressed in the handling of the iron samples in this experiment?

EXPERIMENT 2
Application Lab: Decomposition of an Aqueous Dye*

Background

In this experiment, you will take on the role of a chemical engineer designing a kinetic analysis for the removal of a dye from a waste stream. The reaction you will employ decomposes the dye by reduction. An efficient analytical method is to decompose the dye and measure the decrease in the light absorbed.

CAUTION: During this lab, you will study the decomposition of dye solutions. Be careful—solid iron, though not listed as a toxin, is a fine powder that should be handled carefully to avoid breathing its dust. It should also not be heated, for it can combust.

Procedure

In the report for this experiment, you will be asked how to clean up waste water containing a dye. You will have to decompose all of the dye in 10 min, using a method you devise. It will not, however, be enough for you to dump a large amount of iron into the solution. You will have to justify your method on the basis of some efficiency measures using the data you collect today.

Part I: Formation of Groups

Each group of three will study two different dyes and two different iron samples. Find out from your instructor the cost of the iron reagent. One dye is carmine indigo; the other is methylene blue. Think about how your calculations will require spectrophotometry. What other measurements should you make?

Part II: Design Apparatus and Collect Data for Dye Decomposition (Group Work)

Work together to design a method for monitoring the loss of dye over time, starting from the design in your pre-labs. While doing this, you must pay attention to several factors:

- The reaction is between a solid and a liquid. You must be able to stir or otherwise agitate the solution in a steady fashion, yet also have the solid settle quickly for clean absorbance measurements. Magnetic stirrers will be available.
- You do not know the wavelength at which the dye absorbs the most light. And you must be sure that the starting absorbance of the dye is at most 1.5. Some preliminary experiments may be needed to develop a general plan for the reaction.
- The reaction may start quickly. Therefore, your design should be something you can start in 15 to 60 s.
- Air is known to reverse the decomposition of certain dyes. Therefore, you should minimize contact with air.

---

*The procedure in this lab follows that described in Balko, Barbara A.; Tratnyek, Paul G., *J. Chem. Educ.*, **2001**, *78*, 1661.

When you have a monitoring system designed, draw a sketch of the system in your notebook and show it to your instructor. He or she may suggest improvements or point out flaws. Then collect data for the decomposition of your dye.

You will need to carry out at least eight runs. They should be duplicates of both dyes at two different "loadings" or concentrations of the reactants.

Report

You will need to carry out calculations that describe the dependence of the reaction rate on the concentration of the dye and the amount of iron. Your basic data is the absorbance of light from the dye. Convert this into concentration by assuming that Beer's law applies. Then study the rate law by graphing [dye], ln[dye], and 1/[dye] versus time. Finally, compare the rates when you have added different amounts of iron.

Perform the analyses to determine the rate expression for the mass of iron, m_{Fe}, and for each dye. For your analysis, use the simple expression

$$\text{Rate} = k \, [\text{dye}]^x \, m_{Fe}$$

with the goal of finding x and k for each kind of iron and dye.

Questions to Include in Your Lab Report

1. Report on the utility of the method you developed. How reproducible were the results? Did you notice lag times? How would particle breakup affect your measurements?

2. You know the cost of Megachemco's reactant and how much is required to decompose 95% of 100 gal of 0.01 M dye. Carry out the same calculation for your iron reagents.

Experiment Group **I**

Electrochemical Measurement: Electroplating

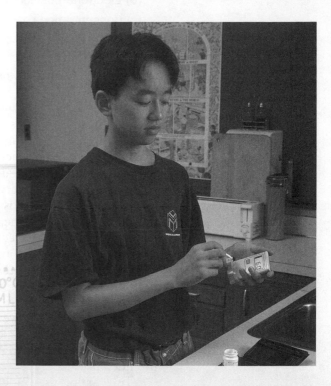

Electrochemical measurement is used in many situations, including glucose monitoring. (David K. Crow/PhotoEdit.)

Purpose　In this experiment group, you will examine both types of electrochemical reactions: chemical reactions that occur and produce electricity (galvanic cells) and chemical reactions that require electricity in order to proceed at all (electrolytic cells). In the first experiment, you will investigate how galvanic cells work and some of the ways in which they are used. Then, using what you have learned about galvanic cells, you will construct and use your own battery to light a bulb. You will also construct an electrode and use it to determine the concentration of an unknown solution. The second experiment in this group focuses on electrolysis reactions, that is, those carried out in an electrolytic cell. After characterizing some qualities of a good electrolysis reaction, you will construct an electrolytic cell and use it to electroplate a thin layer of metal onto another metal electrode.

Schedule of the Labs

Experiment 1: Skill-Building Lab: Electrochemical Measurement and Batteries

- Construct a galvanic cell and determine an activity series of metals by measuring cell potential differences (group work).
- Construct a battery capable of lighting a small bulb (group work).
- Construct a working combination electrode (group work). Use the electrode to:
  (a) Measure $\Delta\mathscr{E}_{cell}$ for standard solutions of Ag^+ needed to prepare a calibration curve of $\Delta\mathscr{E}_{cell}$ versus $\log[Ag^+]$ (group work).
  (b) Determine the concentration of Ag^+ in an unknown solution by comparison with the calibration curve (individual work).

Experiment 2: Application Lab: Electroplating Metals

- Examine the effect of the following factors on current in an electrolysis cell:

  (a) Electrode distance (individual work)

  (b) Electrode position (group work)

  (c) Concentration changes (individual work)

- Devise a plan for and construct an electrolysis cell to electroplate copper (group work).

- Determine Avogadro's number from electrolysis data (individual work).

Scenario

The electrochemical industry produces a vast array of consumer goods, from flashlight and pacemaker batteries to silver-plated jewelry and tableware. These products are possible because of the unique relationship between some chemical reactions—called oxidation–reduction reactions—and electricity. In an oxidation–reduction (redox) reaction, electrons are transferred from one reactant to another. Because electricity is flowing electrons, there is a connection between redox reactions and electricity. Some chemical reactions are used to produce electricity: An example of this is a flashlight battery. In other cases, electricity causes a chemical reaction to occur: An example of this is the electrolysis of water to form hydrogen and oxygen gases.

Batteries can be thought of as containing the components of a redox reaction. With the proper connections, the electricity produced can be used to do work such as lighting a bulb, powering a calculator, or heating the elements used to cook our food. Batteries are produced in various sizes and materials specific to particular needs. We rely on batteries to power our cars, our electronic handheld devices, and personal devices such as hearing aids and pacemakers. All of these objects rely on oxidation–reduction, but the materials and the size of the battery produced depend largely on its intended use.

Two major uses of electrolysis are to refine and protect metals. Electrolytic processes are used to recover aluminum, magnesium, and other active metals from their ores. However, many metals corrode. Corrosion is a destructive chemical process in which metals used in materials such as pipes and girders are attacked by environmental agents, like water and oxygen, and thus become pitted and weakened. This process establishes areas of oxidation and reduction in the metal. In the United States, the cost of corrosion is in excess of tens of millions of dollars per year. The most familiar example of corrosion is rusting. To help protect metals, a thin layer of one metal can be deposited on the surface of another metal, a process known as *electroplating*. The more active metal in the top layer corrodes first, thus protecting the original metal. Electroplating can also be used to provide a decorative metallic surface to other substances, such as chrome-plated automotive and motorcycle parts and silver- and gold-plated tableware.

In this module, you act as the consultant for a large company that produces many different products, all of which rely on electrochemistry. A large section of the company manufactures batteries that will be marketed directly, as well as electroplated metals that will be channeled to other companies that require it as a base for their products. Your task is twofold: (a) to devise and demonstrate a plan to produce a new battery and (b) to devise and demonstrate a plan to electroplate metals.

EXPERIMENT 1
Skill-Building Lab: Electrochemical Measurement and Batteries

Pre-Laboratory Assignment   **Due Before Lab Begins**

NAME: _____

Complete these exercises after reading the experiment but before coming to the laboratory to do it.

1. An electrochemical cell consists of the half-cells $Cu^{2+}|Cu$ and $Ag^+|Ag$.

   (a) Write half-reactions for the reduction of Cu^{2+} to Cu and Ag^+ to Ag. Include the standard reduction potentials, $\mathscr{E}°$.

   (b) Write the half-reactions that occur at the anode and at the cathode. Label these.

   (c) Write the balanced equation for this reaction.

   (d) How many electrons are transferred during this reaction?

   (e) Write the reaction quotient, Q, for this reaction.

2. Sketch and label an electrochemical cell for $Cu^{2+}|Cu$ and $Ag^+|Ag$. Show the direction of flow of electrons in the external circuit for the spontaneous reaction.

3. Study the drawing of the combination electrode you will make in this experiment and answer the following questions.

   (a) What are the components of the reference and of the working electrodes?

   (b) What is the function of the attached string? How must the string be placed so that it functions properly?

4. Use the Nernst equation to calculate $[Ag^+]$ when $[Cu^{2+}] = 0.012$ M and $\Delta \mathscr{E}_{cell} = 0.410$ V.

5. Suppose that you have four solutions containing the metal ions of Fe, Cu, Ag, and Zn. List all possible metal ion pairs.

6. What is the purpose of a calibration curve?

7. You have a stock solution labeled 0.200 M silver nitrate. What volume of stock solution is needed to make up the following dilutions?

   (a) 15.0 mL of 0.075 M silver nitrate

   (b) 15.0 mL of 0.038 M silver nitrate

   (c) 15.0 mL of 0.094 M silver nitrate

8. List two safety precautions that are important to follow when using electricity in an experiment.

EXPERIMENT 1
Skill-Building Lab: Electrochemical Measurement and Batteries

Background

Electrochemistry is the study of the relationship between electricity and oxidation–reduction (redox) reactions. In these reactions an electrical signal such as voltage or current is used to measure chemical amounts (moles) or concentrations (moles per liter). Spontaneous redox reactions generate an electrical potential that is measured by a voltmeter and recorded in volts. This is the basis of a galvanic cell. Nonspontaneous reactions can be forced to occur by applying an external source of electricity to produce a redox reaction. This is an electrolysis cell.

A *spontaneous* reaction is one that occurs without continuous outside help. As with other thermodynamic processes, if a reaction is favored in the forward direction, then it is not favored in the reverse direction. Thermodynamically, spontaneous reactions are favored; nonspontaneous reactions are not favored. Spontaneity reflects the energy of the system. For electrochemical reactions, we can associate the change in the energy of the system with the driving force of the flow of electrons. This force is called *electromotive force* (emf).

In a properly configured system, electrons in a redox reaction will flow through an external wire, and we can use a device to measure their electrical potential or voltage. A spontaneous redox reaction will have a positive cell potential difference, $\Delta\mathscr{E} > 0$. A negative cell potential difference, $\Delta\mathscr{E} < 0$, indicates a nonspontaneous reaction.

Oxidation–Reduction Reactions

A redox reaction is the simultaneous occurrence of two processes: the oxidation process and the reduction process. Oxidation occurs when a chemical species loses or gives up electrons to another chemical species. Reduction occurs when a chemical species receives or gains electrons. The oxidation process provides the electrons necessary for reduction to occur. Therefore, the species oxidized is called the *reducing agent*. The species reduced is called the *oxidizing agent*. Oxidation cannot occur without the corresponding reduction process and vice versa.

Let's look at an example. Fe^{2+} reacts spontaneously with Ce^{4+} according to the equation

$$\overbrace{Fe^{2+} + Ce^{4+} \rightarrow Fe^{3+}}^{-1\ e^-} + Ce^{3+}$$
$$\underbrace{\phantom{Fe^{2+} + Ce^{4+} \rightarrow Fe^{3+} + Ce^{3+}}}_{+1\ e^-}$$

Here we see that the oxidation number of iron changes from $+2$ to $+3$, indicating a loss of electrons. Oxidation numbers will increase in positive value as the chemical species gives up or loses electrons. We recognize that an element in a chemical equation is oxidized whenever that element's oxidation number increases in positive value in the course of the reaction (from reactant to product side of the equation).

The oxidation number for cerium, however, undergoes a decrease in positive value in this reaction, indicating a gain of electrons. Reduction is recognized by the decrease in the positive value of the oxidation number.

In this reaction, Fe^{2+} is oxidized to Fe^{3+} by Ce^{4+}; Ce^{4+} is the oxidizing agent and is reduced to Ce^{3+} by Fe^{2+}; Fe^{2+} is the reducing agent.

The overall redox reaction can be broken down into two half-reactions, one for oxidation and one for reduction:

$$Fe^{2+} \rightarrow Fe^{3+} + e^{-} \qquad \text{oxidation}$$
$$Ce^{4+} + e^{-} \rightarrow Ce^{3+} \qquad \text{reduction}$$

In these half-reactions, we see that Fe^{2+} loses one electron (which appears as a product) and Ce^{3+} gains one electron (which appears as a reactant). In the overall reaction, the number of electrons lost must equal the number of electrons gained. In this case, the requirement is satisfied. When the half-reaction equations are added, the electron terms cancel out to give the overall equation:

$$Fe^{2+} + Ce^{4+} \rightarrow Fe^{3+} + Ce^{3+}$$

Electrochemical Cells

A simple galvanic cell (Figure I-1) consists of two beakers, each with a metal electrode, appropriate wiring, and a voltmeter. The beakers contain aqueous solutions. If a half-reaction involves a pure element, then the electrode is made of that element. Otherwise, an inert electrode is used. The electrode in the beaker where reduction occurs is called the *cathode*, and the electrode in the beaker where oxidation occurs is the *anode*. Reduction always occurs at the cathode; oxidation always occurs at the anode.

Figure I-1   Schematic of a galvanic cell

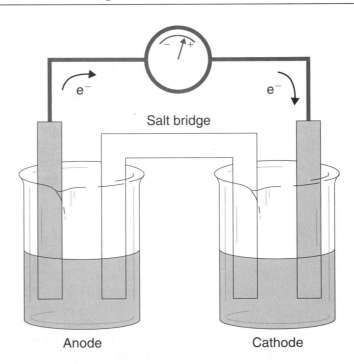

We can monitor what occurs in an electrochemical cell by connecting a meter to the external circuit. Each electrode is connected to the circuit by metal wiring. The meter measures the voltage generated by the redox reaction. The voltage reading will be positive when the electrodes are connected properly for a spontaneous reaction. A spontaneous redox reaction occurs when the species with the higher reduction potential is connected as the cathode. Otherwise, the voltage reading will be negative, which means the cathode and anode have been inverted.

The electrochemical cell for our Fe–Ce example is shown in Figure I-1. The direction of electron flow from the anode to the cathode is also shown. Electrons lost by iron(II) at the anode travel through the wire to the cathode, where they reduce Ce^{4+} to Ce^{3+}. The function of the salt bridge is to maintain electroneutrality in the system as the electrons are transferred from the anode to the cathode during the reaction. The salt bridge contains ions that do not participate in the redox reaction but do migrate in response to the electron flow. As electrons move away from the anode, positive charge builds up around it; as electrons move toward the cathode, it becomes negatively charged. This causes negative ions from the salt bridge to flow toward the anode while positive salt-bridge ions flow toward the cathode.

Galvanic cells similar to that shown in Figure I-1 are easy to assemble and use. However, building such cells with beakers requires a large amount of solution. Your instructor may have alternate materials that use much smaller volumes.

The Activity Series of Metals

Metals have a strong tendency to give up electrons and form positive ions, but the strength of this electron-pushing tendency, or emf, varies from metal to metal. Thus, metals can be ranked according to their ability to give up electrons (that is, according to their oxidation potential). This ranking is called the *activity series* for metals. It is often used to predict whether a particular reaction is likely to occur. As part of this experiment, you will measure the oxidation potentials for the metals selected by your instructor and generate your own mini-activity series.

In today's experiment, you will measure the potential difference between various half-reactions. This will give you experimental values that you can use to rank your metals according to their oxidation potential. Your instructor will tell you what kind of materials to use in the construction of your electrochemical cell. To do this, you will compare electrochemical potential differences in galvanic cells made of different pairs of metal/metal ions. When paired, each metal/metal ion becomes either an oxidation or reduction half-reaction in the galvanic cells you construct.

The Nernst Equation

For measurements taken under standard conditions (1 atm, 1 M solutions), $\Delta\mathscr{E}^{\circ}_{cell}$ measures the standard electric potential difference between the half-cells. For measurements taken under nonstandard conditions (the usual laboratory situation), the Nernst equation is used to calculate $\Delta\mathscr{E}_{cell}$. The Nernst equation gives us the relationship between the overall cell potential difference for a redox reaction ($\Delta\mathscr{E}_{cell}$) and the concentrations of the metal ion solutions.

The Nernst equation is $\Delta\mathscr{E}_{cell} = \Delta\mathscr{E}^{\circ}_{cell} - (RT)/(n\mathscr{F}) \ln Q$, where $\mathscr{F}$ is the Faraday constant, R is the gas constant, n is the number of electrons transferred, and Q is the reaction quotient. Sometimes the natural log (ln) is converted to the base ten log by $\ln x = 2.303 \log x$, and the constants ($RT/\mathscr{F}$) are evaluated using $R = 8.315 \text{ J K}^{-1} \text{ mol}^{-1}$, $T = 298.15 \text{ K}$, and $\mathscr{F} = 96{,}485 \text{ C mol}^{-1}$. Then the equation becomes

$$\Delta\mathscr{E}_{cell} = \Delta\mathscr{E}^{\circ}_{cell} - \left(\frac{0.0592\text{V}}{n}\right) \log Q$$

By manipulating this equation, galvanic cells can be used to determine the concentrations of the metal ions present under conditions other than standard conditions.

Suppose we construct a galvanic cell using silver and tin. The standard half-cell potentials are

$$Ag^+ (aq) + e^- \rightarrow Ag (s) \qquad \mathscr{E}° = 0.7996 \text{ V}$$
$$Sn^{2+} + 2 e^- \rightarrow Sn \qquad \mathscr{E}° = -0.1364 \text{ V}$$

and the spontaneous reaction that occurs is:

$$2 Ag^+ (aq) + Sn (s) \rightarrow 2 Ag (s) + Sn^{2+} (aq) \qquad \Delta\mathscr{E}°_{cell} = 0.9360 \text{ V}$$

For this reaction, the Nernst equation is

$$\Delta\mathscr{E}_{cell} = \Delta\mathscr{E}°_{cell} - \left(\frac{0.0592 \text{ V}}{2}\right) \log \frac{[Sn^{2+}]}{[Ag^+]^2}$$

Now, if we wish to determine the concentration of Ag^+ when $\Delta\mathscr{E}_{cell} = 0.9408$ and $[Sn^{+2}] = 0.010$ M, then $[Ag^+]$ is easily obtained by substituting and solving the equation

$$0.9408 = 0.9360 \text{V} - \frac{0.05916}{2} \log \frac{0.010}{[Ag^+]^2}$$

$$-0.16227 = \log 0.010 - 2 \log [Ag^+]$$

Solving, we get

$$\log[Ag^+] = -0.9189$$
$$[Ag^+] = 0.121 \text{ M}$$

Batteries

Electrical activity in matter was first demonstrated in 1791 when Luigi Galvani observed a continued twitching in frog legs during dissection. He concluded that this electrical activity was a phenomenon associated with living tissues. Several years later, Alessandro Volta demonstrated that electrical activity was also possible in non-living materials. Volta felt a small shock when he touched both ends of a metal stack he had made. "Volta's stack" consisted of alternating zinc and silver metal discs, separated by porous nonconducting material soaked in seawater. A weak electric current flowed when the two ends of the metal stack were connected. Thus, Volta is generally credited with the invention of the battery. Many of today's batteries, such as the carbon–zinc dry cell, rely on the principles developed in the 1860s by France's George Leclanché. Over the years, other metals and solutions have been tested, and today we have many different kinds of batteries designed to meet specific needs.

A battery produces electrical energy by means of a chemical reaction that occurs inside it. Electrons are generated in a redox reaction and passed from the substance oxidized (the anode) to the substance reduced (the cathode). Let's see what this means in terms of a simple example: the zinc–copper battery. Both half-reactions are initially written as reduction half-reactions to correspond with those in the table of reduction potentials. The appropriate half-reactions are

$$Zn^{2+} (aq) + 2 e^- \rightarrow Zn (s) \qquad \mathscr{E}° = -0.76 \text{ V}$$
$$Cu^{2+} (aq) + 2 e^- \rightarrow Cu (s) \qquad \mathscr{E}° = +0.34 \text{ V}$$

We can compare the two $\mathscr{E}°$ values and see that copper has a greater potential to be reduced than does zinc; zinc, then, must be oxidized in this redox pairing, or couple.

The copper half-reaction is written as a reduction half-reaction, but the zinc half-reaction must be reversed to show the oxidation process. To obtain this result, we subtract the two half-reactions:

$$Cu^{2+} (aq) + 2\ e^- \rightarrow Cu\ (s) \qquad\qquad \mathscr{E}° = +0.34\ V$$
$$- (Zn^{2+} (aq) + 2\ e^- \rightarrow Zn\ (s)) \qquad\qquad \mathscr{E}° = -(-0.76\ V)$$

$$\overline{Cu^{2+} (aq) + Zn\ (s) \rightarrow Cu\ (s) + Zn^{2+} (aq) \quad \Delta\mathscr{E}° = 0.34 - (-0.76) = 1.10\ V}$$

The redox couple has a difference in electric potential of 1.10 V.

There are two things we usually want to know about a redox couple: the overall balanced chemical equation for the spontaneous reaction that occurs in a galvanic cell and the $\Delta\mathscr{E}°$ for the standard cell. The overall reaction is obtained in the following manner:

1. Choose the half-reaction with the more positive $\mathscr{E}°$ value as the reduction half-reaction.
2. Subtract (reverse) the other half-reaction to make it an oxidation half-reaction.
3. Adjust both half-reactions so that the electron term has the same coefficient.
4. Combine the two half-reactions.

In the example above, both half-reactions had the same number of electrons, so the $2\ e^-$ terms canceled out. When this is not the case, each half-reaction must be multiplied by a number that will make the electron terms equal, to satisfy the requirement that the number of electrons lost must equal the number of electrons gained. It is also important to note that the $\mathscr{E}°$ values are independent of the multiplication process used with the half-reactions. $\Delta\mathscr{E}°$ depends only on the potential difference between the two metals.

Making Sense of Electrochemistry Terms

Understanding electrochemistry boils down to a few simple concepts.

- Oxidation always occurs at the anode; reduction always occurs at the cathode.
- Given a redox equation, determine which substance is oxidized and which is reduced. Then you can write the half-reaction that occurs at each electrode.

Construction of an Electrode

In the second part of this experiment, you will build your own electrode, calibrate it, and then use it to analyze an unknown. Electrodes are the vehicles by which electrons are shuttled from oxidized to reduced species in a redox reaction. You will make a simple electrode assembly that contains the entire electrochemical cell. This kind of electrode is called a *combination electrode*. One half-cell serves as a reference electrode and consists of a metal in contact with a solution that contains a known concentration of the same metal ion. The electrode you make will use a copper wire immersed in a copper(II) sulfate solution having a concentration that is both known and constant. This reference electrode provides a constant potential for the cell. The special assembly, shown in Figure I-2, also has a second metal electrode—in this case, silver. The silver electrode is called the *working electrode* because it does the work of measuring the concentration of an unknown solution that contains silver ion.

When the $Ag^+|Ag$ half-cell is compared to a reference half-cell, the overall cell potential is given by $\Delta\mathscr{E}_{cell} = \mathscr{E}_{Ag} - \mathscr{E}_{ref}$. You should note that, although the concentration of the copper solution is held constant, it may vary from electrode to electrode so $\mathscr{E}_{ref}$ is different for each electrode. For a given electrode, however, the potential of

Figure I-2   A lab-constructed electrode*

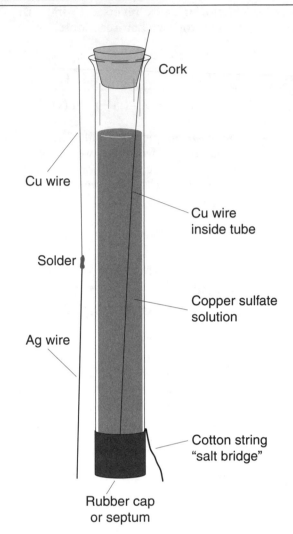

the reference electrode is *constant*. This allows us to focus on only the portion of the Nernst equation for the working electrode—in this case, the silver electrode. The working electrode is always the cathode; the reference electrode is the anode.

Using the Nernst equation, we can evaluate $\mathcal{E}°_{cell}$ for Ag and the relationship simplifies to

$$\Delta\mathcal{E}_{cell} = \Delta\mathcal{E}°_{cell} + \frac{0.05916}{n} \log[Ag^+]$$

where n is 1, the moles of electrons transferred by silver. This equation is of the form $y = mx + b$. The concentration of silver, $[Ag^+]$, can be determined as a function of $\Delta\mathcal{E}_{cell}$ by measuring $\Delta\mathcal{E}_{cell}$ for various standard solutions of Ag^+ and generating a calibration curve. A linear plot results when $\Delta\mathcal{E}_{cell}$ (which corresponds to the y in the general equation) is graphed against $+ \log[Ag^+]$ (which stands in the place of the x in the general equation),

---

*This procedure is modeled on Lisensky, G., and Reynolds, K., *J. Chem. Educ.*, 68, 334–335.

$$\Delta \mathscr{E}_{cell} = \Delta \mathscr{E}^{\circ}_{cell} + 0.05916 \log[Ag^+]$$
$$y \quad = \quad b \quad + \quad (m) \qquad x$$

The slope of the line is 0.05916 and the y-intercept is $\Delta \mathscr{E}^{\circ}_{cell}$. By measuring $\Delta \mathscr{E}_{cell}$ for various known concentrations of silver, a calibration curve can be produced and used to determine $[Ag^+]$ for a solution of Ag^+ of unknown concentration.

Because the copper reference potential may vary from electrode to electrode, you must create a calibration curve for each new electrode you make and use.

Calibration Curves

The equation $\Delta \mathscr{E}_{cell} = \Delta \mathscr{E}^{\circ}_{cell} + 0.05916 \log[Ag^+]$ shows that there is a linear relationship between $\Delta \mathscr{E}_{cell}$ and $\log[Ag^+]$. The voltmeter that produces the measurement of $\Delta \mathscr{E}_{cell}$ can then be calibrated and used to measure $\log[Ag^+]$ for solutions in which $[Ag^+]$ is unknown. The voltmeter is calibrated by measuring $\Delta \mathscr{E}_{cell}$ for several solutions with known concentrations of Ag^+. This generates several pairs of linked $\Delta \mathscr{E}_{cell}$ and $[Ag^+]$ measurements. If we find $\log[Ag^+]$, then we have x, y data in the form of the ordered pair $\log[Ag^+]$, $\Delta \mathscr{E}_{cell}$. With several data points, we can generate a graph of the data, using either Excel or linear regression. Find the y-intercept and the slope of the line and substitute these values into the equation for $\Delta \mathscr{E}_{cell}$ above. This transforms what was a general equation for this type of reaction to a specific equation for your system. To determine an unknown Ag^+ concentration, then, simply measure $\Delta \mathscr{E}_{cell}$ for the unknown solution, substitute this value into the equation from your calibration curve and solve for $[Ag^+]$.

 CAUTION: Observe proper caution in making the electrical connections. In particular, never touch bare wiring that is part of a completed electrical circuit. Many solutions containing metal ions are hazardous and must be handled with care. All solutions should be handled and disposed of properly.

Procedure

Part I: Formation of Groups

In this experiment, you will construct a working battery and a working electrode. For constructing the battery, it is best to work in groups of three. You should plan together how to carry out this part of the experiment but work alone to collect data about the reactivity of metals. Then reconvene as a group to plan and construct your battery.

You will work in these same groups of three to construct an electrode, which will be used to measure the electrochemical potential of solutions with known concentrations of Ag^+. You should discuss how best to divide the work of creating these solutions, measuring $\Delta \mathscr{E}_{cell}$ for each, and making and using the calibration curve. Lastly, you will measure $\Delta \mathscr{E}_{cell}$ for an Ag^+ solution of unknown concentration.

Part II: Construction of an Activity Series of Metals (Group Work)

Several different models of voltmeters may be used for this experiment. Check with your laboratory instructor to be sure that you have the correct model and that it is connected properly to obtain direct current measurements. You must use a direct current (DC) voltmeter and set it to measure volts. To set up the meter for voltage measurements, you will also need two electrical leads with jacks at one end and alligator clips at the other.

This part of the procedure should be done in pairs. Common metals you may use are silver, copper, iron, and zinc. Solutions of silver ion, copper(II) ion, iron(II) ion, and zinc ion will be available.

A number of half-cells can be set up by using beakers (see Figure I-1) or smaller cells. Strips of filter paper soaked in a solution of potassium nitrate or similar salt may be used as salt bridges connecting the two half-cells. Determine the total number of combinations of metals possible. Divide the work of measurement among you.

Dirty connections and electrodes are the main source of error in this experiment. Be certain that the alligator clips are clean before connecting them to the electrodes. You can clean them by pulling some sandpaper between the jaws. When you are confident that all metals are clean and properly placed, connect the alligator clips to any two electrodes. Remember that a galvanic cell is one in which the redox reaction proceeds spontaneously and that a spontaneous redox reaction has $\Delta\mathscr{E} > 0$.

Begin at the highest DC–V setting and gradually lower the setting until you get a reading. If your reading is negative, you have the leads connected to the wrong outlets; reverse them. The meter reading will be positive when the cathode is connected to the $(+)$ outlet and the anode to the $(-)$ outlet of the meter. *Record the voltage and label* which of the two metal electrodes is the cathode and which is the anode for this pairing. Disconnect the alligator clips, *clean them again*, and record a second measurement. If these two measurements do not agree within 5%, repeat the cleaning and measurements again until you obtain reasonable agreement between measurements. Repeat this procedure for all possible pairings of half-cells.

Cleanup.   Clean and return all metals, meters, and leads. Do *not* flush any solutions down the drain. Ask your instructor about how to dispose of them properly.

Part III: Plan and Construction of a Battery (Group Work)

Reconvene as a group to decide which metals to use in constructing your battery. Work together to plan how to make the necessary electrical connections. When your cell is finished, measure its voltage. Record a drawing of your cell; label the components used and the $\Delta\mathscr{E}_{cell}$. When finished, connect your cells together to a single voltmeter. Modify your connections if necessary to obtain a higher $\Delta\mathscr{E}_{cell}$ than you recorded for your individual cells. When you are satisfied with your results, record a drawing as before and note the total $\Delta\mathscr{E}_{cell}$. Test your battery to see if it generates enough electricity to light a bulb. If time permits, repeat this process by connecting additional cells together and obtain the total $\Delta\mathscr{E}_{cell}$ of those combined cells.

Part IV: Constructing a Working Electrode (Group Work)

This part of the procedure is best done in groups of two. The assembly pictured in Figure I-2 contains two electrodes: a $Cu^{2+}|Cu$ reference half-cell and a silver electrode. An electric potential difference will develop between the two electrodes when the silver electrode is put in contact with a solution containing Ag^+, completing the circuit and forming the electrochemical cell.

To construct the electrode assembly,

- Obtain enough materials to make two electrodes.
- Assemble the copper and silver wire electrodes. Each assembly should contain two electrodes: one copper wire and one copper–silver wire (soldered together).
- Insert the copper wire into a length of glass tubing and secure with a stopper or cork.
- Fasten the silver–copper soldered wire to the outside of the tubing with tape.

Fill the reference electrode with 0.1 M $CuSO_4$ solution, insert a string or thread salt bridge, and finally stopper with the septum.

When using this electrode assembly, it is essential that the string, which functions as the electrochemical salt bridge, make contact with the solution inside and outside of the reference electrode.

Part V: Preparation of Standard Solutions and Calibration Curve (Group Work)

Obtain about 25 mL of a standardized solution of Ag^+. Prepare four dilutions such that you end up with four 10-mL solutions, each with a different concentration. The solutions should have concentrations between 0.18 M and 0.05 M. Divide the work evenly among you. Do the calculations for these dilutions now and check them with your partner.

Prepare a table in which to record the concentration of each solution after dilution and the $\Delta \mathscr{E}_{cell}$ measured with your electrode. Obtain a voltmeter and two leads. Make the appropriate connections. Measure $\Delta \mathscr{E}_{cell}$ for each of the solutions.

Part VI: Measurement of an Unknown Concentration of Ag^+ (Individual Work)

Obtain a solution of unknown silver concentration. Record its identifying number or letter. Measure the cell potential difference following the above procedure. Repeat the measurement several times until you are satisfied that your voltage reading is constant. Each member of the group should measure a different unknown solution.

Report

Galvanic Cell Experiments (Individual Work)

1. Write a balanced equation for all ten redox reactions you measured. Circle the metal that served as the cathode in each reaction.

2. Calculate the theoretical value of $\Delta \mathscr{E}_{cell}$ for each reaction by using the Nernst equation, a table of standard reduction potentials, and the actual concentrations of the solutions used.

3. Compare these theoretical cell potential differences with those you *actually* measured and calculate the percentage error for each cell. Comment on possible sources of error in your measurements.

4. Answer the following questions:

   (a) For one pairing that initially gave a negative voltage reading, write the redox equation, identify the anode and cathode, and give the direction of flow of electricity in the external circuit.

   (b) What is the purpose of the KNO_3 salt bridge?

   (c) A galvanic cell cannot generate electricity forever. List two chemical reasons you can think of for why a galvanic cell may go "dead."

Galvanic Cell Experiments (Group Work)

Discuss with your group the best way to determine the ordering of the five metals, from the one that makes the best cathode to the one that is the worst. Include this listing in your lab report.

Construction of Batteries

Include a labeled drawing of your team's connected batteries and the total $\Delta \mathscr{E}_{cell}$. How did you connect the cells in order to increase the overall voltage? Be specific and label your drawing. Was this enough energy to power a light bulb (or some other device)? If not, suggest what you could do to light the bulb.

Construction of an Electrode

1. For your final report, include a sketch of your electrode assembly.

2. Briefly list and discuss any problems you encountered with the procedure.

3. Prepare a graph of $\Delta \mathcal{E}_{cell}$ versus $\log[Ag^+]$. Determine the y-intercept and the slope of the line graphed. Calculate the silver concentration of your unknown solution from its value of $\Delta \mathcal{E}_{cell}$ using the calibration curve equation.

4. Compare the theoretical values for $\Delta \mathcal{E}^{\circ}_{cell}$ and $0.05916/n$ with the values obtained for the y-intercept and slope from your calibration curve.

EXPERIMENT 2
Application Lab: Electroplating Metals

Pre-Laboratory Assignment   **Due Before Lab Begins**

NAME: _____

Complete these exercises after reading the experiment but before coming to the laboratory to do it.

1. What is the difference between a galvanic cell and an electrolytic cell?

2. The following reaction takes place in a *galvanic* cell.

$$Ni\ (s) + Pb^{2+}\ (aq) \rightarrow Pb\ (s) + Ni^{2+}\ (aq)$$

   Identify the anode and cathode in this cell and write the half-reactions that occur at each electrode. Include the $\mathscr{E}°$ values.

3. Now assume that the reaction in Question 2 is part of an *electrolytic* cell and that the reaction we want to occur is

$$Pb\ (s) + Ni^{2+}\ (aq) \rightarrow Ni\ (s) + Pb^{2+}\ (aq)$$

   Identify the anode and cathode and write the half-reactions that occur at each electrode. Will the lead or the nickel electrode increase in mass as the reaction occurs? How can you tell this from the equation?

4. What process occurs at the anode and at the cathode in a galvanic cell? In an electrolytic cell?

5. In the first part of this experiment, you will examine the effect of various parameters on an electrolytic cell. Read through the experiment, think about the process that will occur, and try to predict what will happen when you do the experiment. How do you think the following will affect the current in this cell?

   (a) Electrodes close together versus electrodes far apart

   (b) Wobbly electrodes as compared with rigid ones

   (c) A low versus a high solution concentration

   Give reasons for your answers. (You will have the chance to check your predictions when you actually do the lab. Points will not be taken off for incorrect responses.)

6. (a) Two copper electrodes are used in an electrolytic cell with a solution of copper(II) nitrate. Calculate the mass of copper deposited if a current of 0.50 amperes (A) is run for 45 min.

   (b) Write and label the equations for the reaction that occurs at the anode and at the cathode.

EXPERIMENT 2
Application Lab: Electroplating Metals

Background

In the first lab of this experiment group, you measured the current generated by spontaneous redox reactions in several galvanic cells. In this experiment, you will do the opposite. You will work with nonspontaneous redox reactions as you design and use an electrolytic cell. Whereas a galvanic cell *generates* electricity from spontaneous redox reactions, an electrolytic cell *uses* electricity to cause a nonspontaneous reaction to occur. In both cases, there is a relationship between chemical reactions and electrical energy.

In the galvanic cell experiment, you found that the combination Zn (s) + Cu^{2+} (aq) reacted to produce electricity but that the opposite combination—Cu (s) + Zn^{2+} (aq)—did not react. This agrees with what we predict by looking at the standard reduction potentials of Zn ($\mathscr{E}° = -0.76$ V) and of Cu ($\mathscr{E}° = 0.34$ V). We see from the reduction potentials of both metals that copper has a greater tendency to be reduced (gain electrons) than zinc does. Conversely, zinc has a greater tendency to be oxidized (lose electrons) than copper does. Thus, zinc will give up electrons to copper ions, forming copper atoms.

Cathode reaction (reduction):	Cu^{2+} (aq) + 2 e⁻ → Cu (s)	$\mathscr{E}° = 0.34$ V
Anode reaction (oxidation):	Zn (s) → Zn^{2+} (aq) + 2 e⁻	$\mathscr{E}° = -(-0.76$ V$)$
Spontaneous reaction:	Cu^{2+} (aq) + Zn (s) → Cu (s) + Zn^{2+} (aq)	$\Delta\mathscr{E}° = 1.10$ V

In this reaction, electrons are transferred from zinc to copper ions, and, with the proper connections, electricity flows through the cell. The reaction is spontaneous ($\Delta\mathscr{E}° > 0$). The opposite reaction is nonspontaneous; reversing the direction of the reaction reverses the sign of $\Delta\mathscr{E}°$.

$$Cu\ (s) + Zn^{2+}\ (aq) \rightarrow Cu^{2+}\ (aq) + Zn\ (s) \qquad \Delta\mathscr{E}° = -1.10\ V$$

This second reaction can be made to go in the direction written by using an external source of electricity. In an electrolytic cell, an external current is used to cause electrons to flow in an otherwise nonspontaneous reaction.

Galvanic Versus Electrolytic Cells—Similarities and Differences

Galvanic and electrolytic cells may be difficult to distinguish if you are not careful with your definitions. Both types of cells rely on the same processes and are analyzed in the same manner. First, let's look at the similarities and differences; then we will see how to use this information to analyze electrochemical cells.

Similarities

- Reduction always occurs at the cathode.
- Oxidation always occurs at the anode.
- A species is oxidized when it loses electrons; its oxidation number becomes more positive.
- A species is reduced when it gains electrons; its oxidation number becomes more negative.

- Spontaneity is predicted in cases when the species more likely to be reduced is actually reduced in the reaction.

Differences

- In galvanic cells, electrons are spontaneously transferred in a redox reaction and can be measured by a voltmeter. The cell potential difference is positive.
- In electrolytic cells, an external source of electricity is connected in such a way that electrons are transferred in what, in essence, is a nonspontaneous reaction.

How do you use this information to analyze an electrochemical cell? Generally, you have an equation for the redox reaction, so you can assign oxidation numbers to determine what is oxidized and what is reduced. Then, if the reduced species in the equation has the more positive $\mathscr{E}°$ value, the equation represents a galvanic cell and the reaction will occur spontaneously as written. If not, the equation represents an electrolytic cell.

For example, predict whether tin and aluminum ions will react spontaneously.

$$\text{Sn }(s) + \text{Al}^{3+}\ (aq) \rightarrow \text{Sn}^{2+}\ (aq) + \text{Al }(s)$$

The two half-reactions and standard reduction potentials are

Reduction:	$\text{Al}^{3+} + 3\ e^- \rightarrow \text{Al}^0$	$\mathscr{E}° = -1.66$ V
Oxidation:	$-(\text{Sn}^{2+} + 2\ e^- \rightarrow \text{Sn}^0)$	$\mathscr{E}° = -(-0.14)$ V
Overall reaction:	$3\ \text{Sn }(s) + 2\ \text{Al}^{3+}\ (aq) \rightarrow 3\ \text{Sn}^{2+}\ (aq) + 2\ \text{Al }(s)$	$\Delta\mathscr{E}° = -1.52$ V

We see that the reaction will not occur spontaneously, because $\Delta\mathscr{E}° < 0$.

An alternative solution would compare the two $\mathscr{E}°$ values. The more positive $\mathscr{E}°$ value for tin tells us that tin should be reduced. However, the equation shows that tin is oxidized. The reaction will not occur spontaneously as written.

Another thing you can predict when given a redox reaction is what will physically happen to the electrode during the reaction. In this case, we expect the tin electrode to become smaller as Sn (s) is oxidized to Sn^{2+} (aq); the aluminum electrode should become larger as Al^{3+} (aq) is reduced to Al (s).

Electrolysis

Electrolysis is a chemical process by which electricity is used to effect a chemical change, such as decomposing a compound into its elements. The most familiar example of this is the electrolysis of water into hydrogen and oxygen gases.

$$2\ \text{H}_2\text{O }(l) \xrightarrow{\text{electricity}} 2\ \text{H}_2\ (g) + \text{O}_2\ (g)$$

Molten salts undergo electrolysis as well. Electrolysis of molten NaCl is the commercial process for the production of sodium metal.

$$2\ \text{NaCl }(l) \xrightarrow{\text{electricity}} 2\ \text{Na }(l) + \text{Cl}_2\ (g)$$

Both electrorefining and electroplating are applications of electrolysis. In electrorefining, pure metal is obtained from a metal ore when the metal ore is made the anode in an electrolytic cell. The pure metal plates out on the cathode. One way electroplating is used is to protect metals from corrosion by making them the cathode in an electrolytic cell. Their surface soon becomes coated with a thin protective layer of another metal. A coating of zinc protects galvanized iron or steel. The steel is designated as the cathode in an electrolytic cell and is immersed in a solution of zinc salts, which becomes the anode. The reaction at the cathode is the reduction of zinc:

$$Zn^{2+}\ (aq) + 2\ e^- \rightarrow Zn\ (s)$$

Stoichiometry for Electrochemical Reactions

Electrons are lost in oxidation and gained in reduction. Because the number of electrons lost must equal the number of electrons gained, there is a relationship between the electron flow (electricity) and the amount of matter present. Michael Faraday (1791–1867) recognized that the mass of metal deposited on the cathode of an electrochemical cell was proportional to the quantity of electricity that flowed through the cell. The constant ($\mathscr{F}$) that takes his name (Faraday) is equal to the amount of charge on one mole of electrons:

$$\mathscr{F} = 96{,}500\ \text{coulombs/mole of electrons}$$

Electroplating Copper

In this experiment, you will construct an electrolytic cell with copper electrodes. By comparing the mass of the cathode before and after the electrolysis reaction, you will be able to determine Avogadro's number and an experimental value for $\mathscr{F}$. The information you must acquire from your experiment is the change in mass of the cathode and the total electrical charge that was used. Electrical charge (q) in coulombs is the product of the current, I, in amperes and the time, t, in seconds:

$$q = I \times t$$

Avogadro's number (N_0) is calculated by qM/nmq_e, where M is the molar mass of the metal, n is the number of electrons per atom in the reduction half-reaction, m is the mass of plated metal, and q_e is the charge of an electron.

Faraday's constant, $\mathscr{F}$, is calculated by qM/nm. In this experiment, you will use an optimized method adapted from a procedure developed by Carlos A. Seiglie.[1]

The electrolysis cell consists of two copper pieces (electrodes) and copper(II) sulfate solution, an external power supply, and an ammeter to measure current. The electrode you choose as the cathode should be cleaned and weighed before the electrolysis begins. The initial mass (before electrolysis) is m_i and the final mass (after electrolysis) is m_f. The difference is the mass plated out, m, in the definitions above.

Both the current and the time of electrolysis must be recorded. To obtain better adherence of copper on the cathode, it is necessary to run the electrolysis backward for a short time (t_a) before connecting the chosen copper as the cathode (t_c). The net plating time, t, is proportional to the mass of copper plated out:

$$t = t_c - t_a$$

---

[1]C. A. Seiglie, *J. Chem. Educ.*, **June 2003**, 80, 668–669.

Procedure

Part I: Formation of Groups

This procedure is best done in groups of two. Both individuals are responsible for setting up the equipment needed in this experiment and for measuring some data. In most cases, the data measured by each member of the team involve different solution concentrations and/or amounts. Throughout the experiment (or at the end of the experiment) you should compare your results with those of your partner. This serves as a check of the reproducibility of your data and as an aid in reaching a conclusion.

As a pair, you will investigate some properties of an electrolytic cell. You will collaborate on the design and construction of an electrolytic cell.

Part II: Investigation of an Electrolysis Reaction*

A. Setting Up the Electrolysis Cell (Individual Work)

Obtain two copper electrodes from your lab instructor or from the supply area of your lab. Clean them thoroughly, first with sandpaper or steel wool and then by immersion in dilute HCl for 30 s. Finally, rinse both electrodes with deionized water, dry with a small amount of acetone, and carefully set aside.

Each team member should obtain two beakers. One team member should obtain a 300-mL and a 150-mL beaker. The second team member should obtain a 200-mL and a 100-mL beaker. Between them, the team members should have four different sizes of beakers.

Bend the copper electrodes so that they fit inside the smaller beaker, almost touching the bottom. The bent part of the copper should fit over the side of the beaker so that the electrode remains stationary. Scratch a mark on both electrodes at an equal distance from the bottom of the beaker to serve as a fill-line for the solution when it is added.

Remove the electrodes from the smaller beaker and secure in place in the larger beaker. Add acidified copper(II) sulfate stock solution until the level of solution reaches the fill-line. Now your cell is ready to be wired.

Obtain a DC ammeter and two leads with alligator clips. Connect your electrolysis cell and the ammeter in series and ask your lab instructor to check your wiring before going further. When your wiring is approved by your lab instructor, make the final connection to the DC power supply. DC power is available at your lab table at outlets that require a special three-pronged plug or from a battery pack.

B. The Effect of Electrode Distance (Individual Work)

Position the electrodes so that they are directly opposite one another. Measure and record the distance between the electrodes in this position. For best results, be sure that the electrodes are stabilized—not moving—before you turn on the current. Turn on the ammeter and record the current. Repeat the procedure with solution in the smaller-size beaker. Compare your measurements with those of your partner. Record the electrode distance and the current generated in all four of the different-sized beakers.

C. The Effect of Electrode Position (Individual and Group Work)

Individual Work: With the electrodes in place in the smaller beaker and the ammeter turned on, slowly slide one electrode around the beaker until the two electrodes are almost touching. Record the highest current attained with the electrodes in closest

---

*These experiments are modeled on a procedure reported by Joseph Wang and Carlo Macca, *J. Chem. Educ.* **1996,** *73*(8), 797–800.

proximity. Compare your findings with those of your partner. Record the initial and final currents measured by you and your partner along with the approximate volume of solution in the beakers.

Group Work: Each partner should do the work involved with his or her setup. The other partner will control the ammeter and record current readings. Turn off the ammeter and remove one of the electrodes. Hold the electrode so it is directly opposite and parallel to the other electrode and submerge it in the solution. Turn the ammeter on. Hold the electrode as still as possible and keep the electrode positions parallel. Record the current. Slowly move the electrode forward until the two electrodes are *almost* touching. Record the highest current attained. Repeat the procedure with your partner's setup.

D. The Effect of Concentration Changes (Individual Work)

Reposition the electrodes on the beaker in a fixed position directly opposite one another. Pour out approximately half of the solution from the smaller beaker and replace it with distilled water, filling to the mark on the electrode. Swirl gently to mix; then measure and record the maximum current generated for the diluted solution. Repeat the procedure for a second dilution of the same solution (i.e., pour out half of the solution, fill to the mark with distilled water, and record the maximum current).

Part III: The Electrolysis Reaction (Group Work)

Now that you have investigated some of the possible variations in an electrolytic cell, you are ready to assemble an electrolysis cell to electroplate copper. Discuss as a group the properties you think are important for a good electrolysis cell.

Prepare one piece of copper as the cathode. Use scissors to round off any corners and then wipe clean with a weak organic acid such as vinegar or citric acid. Dry carefully in a warm oven or with a lab tissue. Record the initial mass of the cathode. The prepared piece of copper must be connected as the anode for a short time (about 2 min) and then connected as the cathode for a longer time (about 10 min). Write down what you will do and what you must measure at each step of this procedure. When you are satisfied that you have a complete plan, check with your instructor for approval before doing the actual electrolysis.

Repeat the electrolysis at least one more time. When you are finished with the experiment, make arrangements with at least two other groups to exchange data so that you will have a larger pool of values to average for your final report.

Questions to Answer in Your Report

Group Discussion

Before leaving the lab, discuss the following with the other members of your group.

Part II: Investigation of an Electrolysis Reaction

1. What happens to the voltage when the electrodes are moved relative to one another?
2. What happens to the voltage when the electrodes are close together? When the electrodes are far apart? Is there any relationship or is the difference random?
3. What electrode and solution changes resulted in optimal current?

Part III: The Electrolysis Reaction

1. What information is important to include in your report? What kind of data tables do you need?

2. Obtain data from at least two other lab groups to include in your report.

Individual Report

1. Report your group's conclusions for Part II.

2. *Part III:* Include a sketch of your electrolytic cell, as well as a data table containing your observed measurements and those of at least two other lab groups. Calculate Avogadro's number and Faraday's constant ($\mathscr{F}$) for each set of experimental data. Average your results with those of at least two other lab groups. Report the average value, the standard deviation, and the percentage error for both Avogadro's number and Faraday's constant.

3. Comment on the procedure you followed in Part III. Did it work well or are there some things you would change in your procedure?

Experiment Group **J**

Thermochemistry: Materials and Temperature Control

Sophisticated materials-testing labs work on the same calorimeter principles covered in general chemistry. (Courtesy of John Gadja, Construction Technology Laboratories.)

Purpose   This module is designed to explore the role of heat capacity as a factor in choosing a material with special heat requirements. Heat is a measure of energy transfer into or out of a system and is closely related to temperature. When a system absorbs heat, there is a rise in its temperature; when a system gives off or loses heat, its temperature decreases. In this process, heat is directly related to the temperature change of the system. Thus, if we understand a material's ability to absorb heat, we should be able to control the associated temperature changes.

      Different substances respond differently to heat; some substances require only a little heat to achieve a temperature change, whereas other substances require considerable heat to raise their temperature even one degree. Heat capacity is a measure of this property—it measures the ability of a substance to hold or store heat. Materials with large heat capacities are able to store more heat than materials with smaller heat capacities. Heat capacity is one of the properties that engineers consider when they design systems that must control temperature changes.

Schedule of the Labs

Experiment 1: Skill-Building Lab: Heat Capacity and the Fireproof Safe

- Construct a calorimeter and determine its heat capacity (group work).
- Determine the specific heat for copper (group work).
- Devise a method for measuring the specific heat for other substances, including glass beads, cork, and concrete (group work).
- Choose the best interstitial material for a fireproof safe by comparing the calculated temperature of the safe (after half an hour in the fire) for all the materials measured (individual work).

Experiment 2: Application Lab: The Control of Temperature in Chemical Reactions

- Measure the heat evolved in a chemical reaction (individual work).
- Devise and test a method (using the materials from Lab 1) that will decrease the temperature change for this reaction (group work).

- Use uncertainty analysis to determine the approximate range of acceptable values for the heat of reaction (individual work).
- Compare ΔH with ΔE for this reaction (group work).

Scenario Your company specializes in the design and manufacture of materials used in fireproofing. Although there are many uses for such materials, two of your primary sales targets are companies that produce fireproof containers and those that produce fire-resistant building materials. Fire-resistant or "fireproof" materials do not necessarily protect their interior contents from fire damage indefinitely, but they should protect long enough to allow the real possibility of rescue. In the case of fireproof safes, you need to consider materials that will keep the intense heat of a fire from damaging whatever is inside the safe. The requirements of fire-resistant building materials are similar but more demanding. Fireproof walls and floors must delay the spread of fire to other parts of the building and must also keep the structural support system intact as long as possible, to give occupants and firefighters time to leave the building. Your chief responsibility lies in the research and development of appropriate materials for these two markets.

Your first assignment is to develop a workable blueprint for the manufacture of small, fire-resistant boxes. What do you need to consider? To answer this, you must first bear in mind the purpose of making a safe fireproof. Obviously, you don't want the materials inside to melt or burn. This means you must control the flow of heat between the fire on the outside and the contents of the safe. Most fireproof safes are constructed by putting a smaller box inside a larger box. The filler (or interstitial) material between the two boxes must prevent the heat of the outside fire from damaging the contents inside the safe. The ultimate question, then, is to identify a reasonable interstitial material.

You will have to examine the major factors that govern the amount of heat that can pass through the filler material, as well as other manufacturing considerations such as the cost of such materials. You must also develop a method or technique that can be used to measure the flow of heat into or out of a system. Then you must test all the materials available to your lab group in order to determine an appropriate filler material. You will base your decision on the information you obtained from the calorimetry experiments in the skill-building lab, from your analysis of the factors important to this problem, and by comparing the temperature inside the safe (using several different filler materials) after 30 minutes in a fire. In this application, the filler materials must keep generated heat outside the system.

As preparation for this task, you will construct and use a calorimeter to measure the specific heats of various materials. This will familiarize you with the calorimeter setup and give you practice making the measurements necessary for heat calculations. It will also provide an experience of heat measurement different from Experiment 2 in this experiment group. In Experiment 2, you will measure the heat evolved in a chemical reaction that takes place in aqueous solution.

The second experiment mimics the heat produced in runaway chemical fires such as the combustion of airplane fuel in the World Trade Center disaster. Although the WTC buildings were initially weakened by the impacts of the planes, it was the resulting fires that caused the structures to collapse. City fire codes require fireproofing in high-rise buildings that will contain a fire for approximately two hours. This generally allows time to evacuate occupants, salvage possessions, or quench the fire. In the case of the WTC, however, evidence ultimately showed that the impacts had knocked some of the fireproofing material off of the steel supports.

Your second task is to find a material that will absorb enough heat to fireproof structural materials. You will apply what you have learned about heat capacity to identify a substance capable of decreasing the heat of a mock runaway reaction.

EXPERIMENT 1
Skill-Building Lab: Heat Capacity and the Fireproof Safe

Pre-Laboratory Assignment   **Due Before Lab Begins**

NAME: _____

Complete these exercises after reading the experiment but before coming to the
laboratory to do it.

1. One of the factors engineers must consider is the validity of any assumptions
   made while they perform their jobs. In this experiment, several temperature
   assumptions are made. List two of these assumptions and explain why you think
   they are reasonable assumptions to make.

2. Two common mistakes made in this experiment are (1) loss of heat when the
   hot substance is transferred to the cold water and (2) failure to attain a constant
   final temperature. Discuss the steps you will take to avoid these mistakes.

3. During an experiment, you perform many different procedures and the final
   results of your experiment depend largely on how well you carry out these
   procedures as well as on the design of your lab apparatus. Decide whether the
   following conditions will have a large or a small effect on your experimental
   results. Give a reason for your answer.

   (a) The calorimeter lid has a large hole in it.

   (b) Hot water is transferred along with the metal into the calorimeter water.

   (c) The initial temperature of the cold water is 28 °C.

   (d) You forget to put the lid on the calorimeter.

(e) You use an alcohol thermometer instead of a mercury thermometer.

(f) You record the final temperature before thermal equilibrium is established.

4. Suppose a hot (89.5 °C) piece of copper metal ($c_s = 0.385$ J g^{-1} K^{-1}) with a mass of 2.55 g is put into 50.0 g of water ($c_s = 4.184$ J g^{-1} K^{-1}). Calculate the final temperature of the water if its initial temperature is 25.7 °C.

5. A fireproof safe is made of two thin metal boxes, one inside the other, with equal spacing between each pair of adjacent surfaces. Calculate the volume *between* the two boxes given the following dimensions. Include units.

      Smaller box: 25 dm × 20 dm × 15 dm
      Larger box: 32 dm × 27 dm × 22 dm

6. Predict which material will be best for the fireproof safe. Give your reasons.

7. Suppose you spill the boiling water (used in this experiment to heat irregular solid substances) on your arm. What first aid procedure would you use to control the damage to the burned arm?

8. What is the general method used in your lab to heat water? List two safety precautions you must heed during the heating process.

EXPERIMENT 1
Skill-Building Lab: Heat Capacity and the Fireproof Safe

Background

Heat capacity measures the amount of heat needed to raise the temperature of a substance one degree. In other words, heat capacity measures the ratio of heat to temperature change, $C = q/\Delta T$, where C is the heat capacity, q is the heat, and $\Delta T = T_{final} - T_{initial} = T_f - T_i$. When we examine the relationship given by this formula, we see that the magnitude of C depends on the magnitude of the ratio $q/\Delta T$. C will be large when $q/\Delta T$ is large; C will be small when $q/\Delta T$ is small. It is logical that ΔT must be kept small to protect materials from fire damage, that is, the large amounts of heat produced by fires. This implies a large value of the ratio $q/\Delta T$ and suggests that the research and development personnel might do well to consider candidate materials with large heat capacities. The heat produced in various types of fires is known or can be measured, and ΔT is determined by the nature of the materials that must be protected from fire damage.

Heat capacity is an important property to industries whose livelihood depends on the ability to control heat transfer. The heat capacity of an object depends on the size (mass) of the object. It is an extensive property. For example, 2000 kg of water has a greater heat capacity than 20 kg of water. In other words, 2000 kg of water can store more heat than 20 kg of water. To directly compare the heat-storing capability of two or more substances, we use **specific heat capacity** (c_s), which is the heat capacity *per gram* of a substance, $c_s = q/(m\,\Delta T)$. The specific heat capacity is used in the calculation of heat,

$$q = mc_s(T_f - T_i) = mc_s\,\Delta T$$

where m is the mass of the substance in grams and c_s has units of J g^{-1} K^{-1}.

For mole amounts, we use the **molar heat capacity** (c_p). Molar heat capacity is the amount of heat needed to raise the temperature of one *mole* of substance one degree. Then the molar heat capacity is $c_p = q/(n\,\Delta T)$ and heat is calculated as $q = nc_p\Delta T$, where n = number of moles of a substance and c_p is the molar heat capacity with units of J mol^{-1} K^{-1}.

When expressing the absolute temperature of a system, we usually *must* convert to the Kelvin scale. However, when measuring temperature **change,** we may use either the Kelvin or the Celsius scales. The Kelvin and Celsius temperature scales are different, but, because the size of the degree unit in these two scales is the same, ΔT has the same numeric value in both temperature scales. A temperature change from 32.4 to 26.2 °C gives $\Delta T = -6.2$ °C. The corresponding change on the Kelvin scale is from 305.6 to 299.4 K; $\Delta T = -6.2$ K.

In this experiment, you will learn how to determine the heat capacity of a metal (copper) that may be useful in a safe. You will then examine the heat capacities of several other substances that might be good filler materials for a safe: glass beads, cork, and concrete. The shapes of some of these materials may require you to explore other methods of heating them to attain the initial high temperature. Not all of these materials will fit inside a test tube. Although we might consider direct submersion into the hot water, this introduces some degree of error if any of the hot water is transferred to the cold water container along with the substance. In addition, direct submersion is a poor choice for either the cork, which is porous and absorbs water, or the concrete, which is a mixture and tends to fall apart in the hot water bath. A heat-resistant plastic wrap or bag would be a good alternative to the test tube we will use for metal pellets. The plastic material allows close thermal contact between the solid and the hot water while preventing the solid from becoming wet.

By the end of this experiment, you will be asked to make a choice of filler materials based on your experimental results for the specific heat capacity of the candidate materials. The choice may depend on a variety of properties that you didn't test. Are there other factors you should consider in your choice of material? Suppose the only important property to consider is thermal conductivity. Thermal conductivity measures the transfer of heat by conduction across a solid. How do our materials compare in their ability to conduct heat? Instead of doing the experiment, we will look up these values in *The CRC Handbook of Chemistry and Physics*. The thermal conductivity of cork is 0.04 cal s^{-1} cm^{-2}. The thermal conductivity values of concrete, copper, and glass are 1.4 cal s^{-1} cm^{-2}, 400 cal s^{-1} cm^{-2}, and 28 cal s^{-1} cm^{-2}, respectively. If you consider only thermal conductivity for the safe, which one of these materials do you think would be the best to use? At the end of this experiment, you will be able to check on the validity of your guess.

A typical design for a fireproof safe is shown in Figure J-1. The inner box has the dimensions height = 0.100 m, depth = 0.120 m, and width = 0.240 m. The outer box has the dimensions height = 0.134 m, depth = 0.154 m, and width = 0.274 m, making the space between adjacent surfaces 0.017 m. A material is to be selected for the interstitial space between the two metal boxes such that the following design specifications are met.

Design Specifications

1. The interior of the safe is not to exceed 177 °C when exposed to a fire of temperature 954 °C for half an hour.
2. The safe is to be portable, so that an average person can carry it. Minimum weight is desired.
3. The safe is to be a consumer product. Cost is to be minimized.

Using an equation we will give you later, you can calculate the inner temperature of the safe after half an hour of exposure to the fire temperature for each interstitial material. The temperature must *remain less than 177 °C* for the safe to be classified as fireproof. You will have to calculate the mass of each material needed to fill the interstitial space in the safe. You can do this by calculating the volume between the inner and outer boxes and by measuring the density of each substance used in this lab. Then, mass equals the product of density and volume. For this calculation, we will assume that the geometry of the fireproof safe is fixed for all materials. You can determine the cost of each material needed to fill the interstitial space if you know the cost per unit mass. Finally, you will choose one of the materials you tested as best suited for the fireproof safe.

Figure J-1

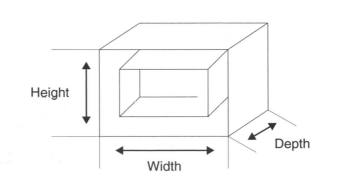

Calorimetry

Heat capacity values can be experimentally determined using calorimetry methods. Calorimetry experiments use initial and final temperature measurements to determine the heat generated or absorbed by some change. The change may be as simple as the heat transfer between a hot piece of metal and cold water, or it may involve measuring the heat of a chemical reaction. To determine ΔT in a calorimetry experiment, it is necessary to control the heat transfer by isolating the system of interest in a well-insulated container called a *calorimeter*.

In thermodynamics, we define the universe as consisting of two parts: (1) a particular system of interest (usually just called the *system*) and (2) everything else outside the system (usually called the *surroundings*). Energy can be transferred between the system and its surroundings, but the total amount of energy distributed between the two remains constant. In other words, $\Delta E_{univ} = \Delta E_{sys} + \Delta E_{surr} = 0$. For this experiment, the surroundings will be the calorimeter; the system, then, is the contents of the calorimeter. We are only concerned with thermal energy monitored by the variable heat, q, so $q_{univ} = 0 = q_{sys} + q_{surr}$.

Determining the Heat Capacity of the Calorimeter

Calorimeters can be constructed using a variety of designs. The Styrofoam® cup containing your hot morning coffee is a type of calorimeter designed to minimize the transfer of heat from the coffee to the surrounding air. The calorimeter you will use in this experiment will have a lid so that essentially all of the heat transferred stays inside the calorimeter. Because the calorimeter is the only component of the surroundings, we assume that $q_{surr} = q_{cal}$.

The amount of heat actually absorbed by the calorimeter can be determined experimentally by measuring the heat transfer that occurs when hot water is added to cold water inside the calorimeter. Theoretically, the heat lost by the hot water should equal the heat gained by the cold water. This is the law of conservation of energy. Any discrepancy in the two values is due to the heat absorbed by the calorimeter, which includes the thermometer and stirring bar. The amount of heat absorbed by the calorimeter is related to the heat capacity of the calorimeter and is expected to be the same for similar systems. The heat capacity of the calorimeter is symbolized as C_{cal}. C_{cal} has units of $J\ K^{-1}$.

The overall system is adiabatic ($q_{univ} = 0$), so the calorimeter constant is calculated by

$$q_{univ} = 0 = q_{sys} + q_{surr}$$
$$q_{surr} = q_{cal}$$
$$q_{sys} = q_{cw} + q_{hw}$$
$$q_{univ} = 0 = q_{cw} + q_{hw} + q_{cal}$$
$$q_{cw} = c_{cw}m_{cw}\Delta t_{cw}$$
$$q_{hw} = c_{hw}m_{hw}\Delta t_{hw}$$
$$q_{cal} = -(q_{cw} + q_{hw}) = C\Delta t_{cal}$$

where q_{cw} = heat change for the cold water
q_{hw} = heat change for the hot water
q_{cal} = heat change for the calorimeter
Δt_{cal} = Δt of initial contents of calorimeter.

The specific heat of water is $c_{cw} \approx c_{hw} \approx 4.184\ J\ g^{-1}\ K^{-1}$.

To determine q_{cal}, you need to evaluate q_{cw} and q_{hw}. For the same calorimeter, C_{cal} may be considered constant when used with comparable masses of water. However, ΔT will most likely vary from experiment to experiment, so q_{cal} must be recalculated for each experiment.

Determining the Specific Heat Capacity of Substances

In this experiment, you will also measure the temperature of cold water before and after some hot metal is added to it. When two substances having different temperatures are placed into close physical contact with one another, heat will flow from the hotter substance into the colder substance until both substances arrive at the same temperature. They will attain *thermal equilibrium*. A simple example of this is the immersion of a hot metal into a beaker of cold water. The transfer of heat from the metal to the surrounding water and beaker can be followed by monitoring the temperature change of the water. The water temperature measured when the metal–water system has reached thermal equilibrium is the final temperature for both substances. Measurement of this temperature change allows one to calculate the heat transferred from the metal to the cold water and the calorimeter. Heat is calculated as the product of mass, specific heat capacity, and temperature change: $q = mc_s\Delta t$.

As the system approaches thermal equilibrium, the heat given off by the metal and the heat absorbed by the surrounding water and beaker are equal but are given opposite signs by convention. In thermodynamics, whatever leaves the system is considered a negative quantity while that which is absorbed by or added to the system is considered to be positive in sign. In this experiment, heat leaves the metal and is absorbed by the cold water and the calorimeter.

Because your overall system is adiabatic, we can write

$$q_{univ} = 0 = q_{sys} + q_{surr}$$
$$q_{sys} = q_{cw} + q_{metal}$$
$$q_{surr} = q_{cal}$$
$$q_{cw} + q_{metal} + q_{cal} = 0$$
$$q_{cw} + q_{cal} = -q_{metal}$$
$$(mc_s\Delta t)_{cw} + C\Delta t_{cal} = -(mc_s\Delta t)_{metal}$$

Note that in this case, the values for Δt will be different for the metal and the cold water because these are different substances.

Substances attain thermal equilibrium when their temperatures become equal after some period of close physical contact. A solid substance can attain thermal equilibrium with a second substance (such as water) in many different ways. In the first part of this experiment, the solid substance is a metal and is not soluble in water. To bring the metal to the temperature of the hot water, you need an experimental design that will keep the metal dry and allow you to transfer the heated metal quickly (and safely) to the beaker containing cold water. Large pieces of metal could reasonably be contained in a heat-resistant zip-lock plastic bag. Small pellets of metal could be heated inside a test tube immersed in hot water. With the passage of 7 to 10 min, it is reasonable to assume that the heat from the water bath has been transferred to all of the metal. This is generally a valid assumption, and it has the additional advantage of ensuring that the final temperature measured is due solely to the transfer of heat from the metal to the cool water and not from any drops of water from the hot water bath. By initially keeping the metal sample dry, no water from the hot water bath is transferred into the calorimeter beaker along with the heated metal.

The setup for this experiment is straightforward. It involves two containers of water, one that is used to heat a known mass of the metal to some measured, elevated temperature, and a second, cooler sample of water in the calorimeter into which the hot metal will be put. The amount of water in the first beaker is unimportant because it is used merely to heat the metal. It is crucial, however, to measure the mass of water in the *second* beaker because this mass enters into the calculation of heat. The initial temperature of the hot metal will be assumed to be the same as that of the hot, boiling water. The initial temperature of the cold water will be measured before the addition of the hot metal. The final temperature of both the metal and the cold water will be the same—the highest water temperature attained after the addition of the hot metal to the cold water. With these measurements, you can calculate the specific heat of the metal.

Throughout the experiment and the calculations, remember three important points:

1. The major assumption is that no heat flows out of the calorimeter.
2. Neither the heat capacity nor the specific heat of any component changes either with temperature or with time.
3. Heat absorbed is a positive quantity; heat given off is a negative quantity. Or, if something "gives off" heat, then it "absorbs" a negative amount of heat.

Calorimetry experiments are relatively simple to perform, but you should be careful to avoid some common errors. One source of error is in measurement. In calorimetry, we measure mass and temperature. As the calculation of heat (q) depends strongly on the quality of these two measurements, you should take great care to obtain the most accurate measurements possible. The more difficult of the two measurements is temperature or, more precisely, the final temperature. After adding the hot substance to the cold water, you should continue taking temperature readings every 15 s until there is no change in temperature for three successive readings. Try not to be impatient; stopping too soon will greatly affect your final results.

A second source of error lies in not performing the steps in the procedure quickly and cleanly. In this experiment, you have a colder substance (generally water) and a warmer substance (for example, water, metal pellets, a chunk of concrete). You need the initial temperature of each; then you need a final temperature after the warmer substance has been added to the colder water. If you add warm water to cold water, the transfer and final temperature measurement are straightforward to obtain. However, you may have to use some imagination in the heating and transfer of some of the other materials. Is the material you are heating actually at the temperature of the hot water? Have you allowed it to come to thermal equilibrium? How much heat are you losing in the transfer from the hot water bath to the cold water in the calorimeter? Is it negligible? How can you decrease the heat loss? Questions such as these should be discussed with your lab partner so that you may optimize the quality of your lab results.

 CAUTION: This experiment necessitates the use of hot water. The two most common methods of heating water in the lab use either a bunsen burner or a hot plate. If your lab uses open flames, such as those produced by a bunsen burner, be careful that no part of your body (including your hair) or your clothing comes into contact with the fire. You must wear goggles at all times in the lab. You must also be careful that other people in your lab are not working with flammable chemicals. Flammable chemicals must not be opened or used in the vicinity of an open flame.

If your lab uses hot plates for heating purposes, follow the directions of your lab instructor about the location and usage of the hot plates. Most hot plates are equipped with a light that is on when the hot plate is on. Hot plates may appear the same to you whether they are turned on (and hot) or turned off (and cold). Observe proper precautions when working in the vicinity of the hot plate.

Procedure

Part I: Formation of Groups

For group discussion and data analysis purposes, this experiment is best done in groups of two. Your group should collaborate on the construction of the calorimeter and all measurements needed to calculate specific heat values. You will construct and use a calorimeter, and then determine the calorimeter constant and the specific heat capacity of copper. In Parts IV, V, and VI you will divide the work to be done between you.

Engineers do many experiments during the course of a day, a week, or a year. They must be prepared to discuss, analyze, or report on any one of these experiments at any time. Part of their job is to carefully record the laboratory setup, conditions, assumptions, and all data that might help them fulfill these responsibilities. Measurements, changes in procedure, accidental spillage, and all other details that may influence experimental results should be recorded in your notebook so that everyone in the group can know what happened. Take time to prepare your notebook; indicate the method used, sketch a picture, and keep track of problems and difficulties so that you can avoid them in the future.

Begin now by preparing a table that specifies the measurements you will contribute to the group. Your notebook should include the masses and temperatures needed for the calculation of the calorimeter constant and the specific heat capacity of different materials.

Part II: Construction of a Calorimeter (Group Work)

There are several kinds of calorimeters you may use in this experiment. One is a simple Styrofoam cup with a foil cover. It has the advantage that it is easy to construct and use. However, the light weight of the Styrofoam means that you must be especially careful to avoid tipping it over.

To use Styrofoam cups, obtain two cups to nest one inside the other and a 3 × 3 in. square of foil to form a cover for the nested cups. Carefully mold the foil to the top of the cup assembly so that it can be removed and replaced easily. Make a small hole in the center of the foil just large enough to accommodate a thermometer. The nested Styrofoam cups can be stabilized by resting them inside a glass beaker.

A sturdier calorimeter may be made by placing a 250-mL beaker inside a larger beaker or inside a special metal can. The 250-mL beaker is used as the inner cup of the calorimeter. If the outer part of the calorimeter is a larger beaker, you can use paper toweling as insulation between the two beakers. There are also metal cans available to use as the outer part of the calorimeter. In this case, the 250-mL beaker is supported by a metal ring that fits inside the can. When properly assembled, the beaker is suspended by the insulating metal ring so that no part of the beaker touches the can. Further insulation can be provided by wrapping the inner beaker with paper towels. A lid can be fashioned from a 5 × 5 in. square of cardboard. Cut a small hole in the center of the cardboard for the thermometer. Wrap a rubber band around the thermometer to position it so that it does not touch the bottom of the calorimeter beaker.

Yet another calorimeter is a small thermos. In fact, any insulated container makes an acceptable calorimeter if it minimizes the transfer of heat between the inside reaction container and the atmosphere. Follow the direction of your laboratory instructor in your choice of calorimeter materials.

Part III: Measurement of the Heat Capacity of the Calorimeter (Group Work)

Weigh out two samples of deionized water. An appropriate amount of water for this experiment is about 50 g. The mass can be any value within 10% of this. Record all masses to a precision of ±0.01 g. Put one water sample into the calorimeter beaker and measure its initial temperature. Measure the thermometer readings at eye level and carefully estimate the temperature to within ±0.2 °C. Heat the second sample of water until its temperature is about 10 to 15° higher than that of the cold water.

Record the initial temperature of the warm water. When you are ready, quickly pour the warm water into the calorimeter beaker with the cooler water, replace the lid, and monitor the temperature as it changes. Record the temperature readings every 15 s for 3 to 5 min. Note the highest temperature attained by the water mixture in the calorimeter. The system is assumed to be at thermal equilibrium when the change of temperature over time becomes very small. There is only one final temperature; it is the same for both the hot and the cold water.

Repeat for a total of two trials.

During the laboratory period it is important that you detect any mistakes in procedure or technique before going on to the next part of the experiment. Calculate the heat capacity of your calorimeter for each trial. The two values should be within 5% of each other. If not, repeat the determination until the values are consistent. Use an averaged value of C_{cal} for the remainder of the calculations.

Part IV: Obtain the Specific Heat Capacity of Copper (Group Work)
Preparation of the Calorimeter Water

Place about 100 g of deionized water into the calorimeter beaker. (*Record* the precise mass in your laboratory notebook.)

Record the temperature of the water to ±0.2 °C. This temperature is the *initial temperature* of the water in your calculations.

Heating the Sample

Half-fill a test tube with an accurately measured mass of copper metal pellets. Heat the test tube and metal contents to thermal equilibrium in a hot water bath, being careful to keep the metal dry. However, to guarantee that all of the metal has reached thermal equilibrium, be careful to keep the level of the water surrounding the test tube above the contents of the test tube. Allow about 7 to 10 min for thermal equilibrium to be established. Then record the temperature of the water bath to the same degree of precision as you did for the cooler water in the calorimeter. This will be the initial temperature of the metal in your calculations.

Pour the hot metal into the water in the calorimeter. To do this transfer quickly but efficiently, work together to take off and replace the calorimeter lid. Carefully mix the contents of the calorimeter while you watch the thermometer. Note and record the highest temperature attained (to ±0.2 °C). This number is the *final temperature* of the water in the calorimeter, the calorimeter itself, *and* the metal, as all three are in thermal contact. Perform two experimental trials.

Part V: Determining the Specific Heat Capacity of Cork, Glass, and Concrete (Group Work)

Plan a procedure for the measurement of specific heat with samples of cork, glass, and concrete, or other materials supplied by your instructor. A different group member should set up and direct each of these. Do two trials for each substance. These substances require a full 15 min in the hot water bath for thermal equilibrium to be reached. Decreasing this time will result in very poor calculated values for specific heat capacity.

Part VI: Determining Density

It is also necessary to determine the *density* of each substance, as density is one of the parameters included in the calculation of the inside temperature of a safe surrounded by a raging fire. Determine the densities of all the materials for which you determined specific heat capacity in this experiment.

Report

1. *Copper:* Report your two measured values together with at least four other values obtained by other members of your class. Find the average value for the specific heat of copper. Calculate the percentage of error between your averaged result and the reported value found in the *CRC Handbook of Chemistry and Physics* or your textbook.

2. Include in your lab report a calculation of the uncertainty propagated through the calculation of specific heat capacity for copper.

3. Calculate the specific heat capacity for all the substances measured in Parts V and VI. Report the averaged value obtained from your two trials for each substance.

4. Calculate the temperature inside a fireproof safe (T_{safe}) during a fire, using each of the candidates for filler material. An estimation of T_{safe} can be obtained from

$$T_{safe} = T_{fire} - (T_{fire} - T_{safe, initial})e^{\left(-\frac{hA_s}{c_s m}t\right)}$$

where $T_{fire} = 954\ °C$, $T_{safe, initial}$ = room temperature, and h is the heat transfer coefficient and is assumed constant for the three materials. [It is calculated to be $6\ J\ (s^{-1}\ m^{-2}\ K^{-1})$.] A_s represents the exterior surface area of the safe in square meters, t represents time in seconds, and c_s and m refer to the specific heat and mass (in grams) of the interstitial material of the safe.

5. Prepare a table that includes the following data for all measured substances: specific heat capacity, density, thermal conductivity, mass to fill interstitial space in safe, and T_{safe}.

Group Discussion

Discuss the following together and include your answers in the final report of each group member.

1. The equation used to calculate the inside safe temperature contains c_s as a variable. How does the material with the highest c_s value rate when the inside temperature is calculated? (Refer to your table of heat capacity values for these substances.)

2. Do your experimental findings support the statement, "Substances with large specific heat capacity values are good 'heat-storing' substances"? Why or why not?

3. Compare the half-hour safe temperature calculated using each of the materials. The inside temperature cannot exceed 177 °C for the safe to be considered fireproof. The substances that meet or come closest to meeting this specification are possible filler materials. List them.

4. Which property do you think is the most important in determining a good interstitial material for a fireproof safe? Why?

5. Which material is best suited as the filler material for the fireproof safe? Defend your choice. Did you include cost in your decision?

6. What happened to the thermal conductivity comparison you made (that is, how did that parameter enter into your choice for the best interstitial material)? Did your intuition direct you to choose the lowest conductivity material (that is, the cork)? Did your experiments and calculations support your intuition?

7. As a final check on your results, check the weight of a commercially available fireproof safe. Would you say that the manufacturer came to a conclusion similar to yours for the interstitial material of the safe?

EXPERIMENT 2
Application Lab: The Control of Temperature in Chemical Reactions

NAME: _____

Complete these exercises after reading the experiment but before coming to the laboratory to do it.

1. Which of the substances from the skill-building lab had the highest heat capacity? Which had the lowest heat capacity? Which of these substances is better at storing heat? Defend your answer by calculating the heat absorbed by 5.00 g of each of these substances, assuming $\Delta T = 30.00$ °C. Use your experimental values for specific heat capacity.

2. List some possible sources of error in a calorimetry experiment and detail how your group might have had better results for the substances you measured in the skill-building lab.

3. Explain what is meant by the term *adiabatic*. Name two assumptions concerning heat transfer that are validated because the system is adiabatic.

4. Calculate the change in internal energy (ΔE) that occurs in the combustion of 5.00 g of methanol, CH_3OH, at a temperature of 298 K. Methanol burns according to

$$CH_3OH \ (g) + \frac{3}{2} \ O_2 \ (g) \rightarrow CO_2 \ (g) + 2 \ H_2O \ (g) \qquad \Delta H = -676.5 \ kJ$$

5. Calculate the moles of H_2O_2 in 50.00 g of a 3% by mass solution of H_2O_2.

6. Because it is important for you to measure temperatures accurately in this experiment, you may be using mercury thermometers. Why must you not use the thermometer as a stirrer in your calorimeter?

EXPERIMENT 2
Application Lab: The Control of Temperature in Chemical Reactions

Background

The first law of thermodynamics tells us that the total amount of energy in the universe is constant. We usually think of this as the law of conservation of energy. The interesting thing about energy is that it can take many different forms—heat, light, and electricity are all ways in which energy is manifested. Because energy can be transferred from substance to substance using many different vehicles, it can be difficult to keep track of the energy involved in chemical processes.

In a chemical reaction, the system of interest is the reaction itself and the surroundings are usually taken to mean the immediate surroundings, including the container and the surrounding air; in short, anything that touches the system. Ordinary chemical reactions allow us to measure some interesting thermodynamic quantities. For example, the internal energy of the system changes with each bond that is broken or formed. The change in internal energy (ΔE) is the sum of two terms, heat and work:

$$\Delta E = q + w$$

For experiments performed under ordinary laboratory conditions, the pressure is assumed to be constant. This is shown in measurements of heat by a subscript p, signifying heat at constant pressure (q_p). Heat at constant pressure is the same as the change in enthalpy, ΔH:

$$q_p = \Delta H$$

For systems under normal laboratory conditions of pressure, then, we can measure ΔH quite simply by measuring q_p.

The change in internal energy (ΔE) is

$$\Delta E = q_p + w = \Delta H + w = \Delta H - P\,\Delta V$$

Here the work term simplifies to the volume expansion of a gas against the constant pressure of the atmosphere.

For chemical reactions that involve no expansion of gases, $w = 0$ and

$$\Delta E = \Delta H = q_p$$

For chemical reactions that generate a number of moles of gas that differs from that originally present in the reactants, however, the work term must be evaluated. If we assume that any gas generated by the chemical reaction behaves as an ideal gas, then V can be evaluated by making appropriate substitutions using the ideal gas equation,

$$V = \frac{nRT}{P}$$

Then

$$P\,\Delta V = P\,\Delta\left(\frac{nRT}{P}\right) = RT\,\Delta n$$

where T and R (8.314 J mol^{-1} K^{-1}) are constant, and Δn = number of moles of gas (product) − number of moles of gas (reactants). This substitution allows a calculation of ΔE given ΔH, the temperature, and the chemical equation for the reaction.

Heat

Heat is a means by which energy can be transferred between a system and its surroundings. Some reactions require the addition of heat. These are called **endothermic** reactions. As endothermic reactions absorb heat, $q_{rxn} = \Delta H_{rxn} > 0$. Reactions that liberate heat to their surroundings are called **exothermic** reactions. In an exothermic reaction, $q_{rxn} = \Delta H_{rxn} < 0$. Although heat and temperature are not the same thing, there is a direct correlation between them. As heat flows into an object, the temperature generally increases; as heat flows out of an object, the temperature decreases. Hence, heat transfer can be monitored and measured by watching the temperature.

Thermochemistry is a branch of chemistry that studies heat transfer. For heat transfer to occur there must be thermal contact between the substances in the system. To measure the heat transferred during a chemical reaction, it is imperative that we prevent the heat from escaping or diffusing into the surrounding atmosphere before a proper measurement can be obtained. This is accomplished by using a calorimeter.

In this experiment, you will put a weighed amount of H_2O_2 solution into a calorimeter and measure the heat that is given off by the decomposition reaction of H_2O_2 occurring in the solution. You can assume that all of the heat transferred in the reaction stays inside the calorimeter (that is, assume adiabatic conditions). The H_2O_2 solution contains a very small amount of H_2O_2 (1–5% by mass) and a predominant amount of H_2O (99–95% by mass). The heat liberated by the decomposition of H_2O_2 is absorbed by the solution water and by the calorimeter.

The Decomposition Reaction

We often use a dilute solution of hydrogen peroxide, H_2O_2, to thoroughly cleanse cuts and abrasions. It is a familiar first aid remedy. When H_2O_2 contacts your skin, it bubbles vigorously, cleansing away surface grit. Warmth is a secondary sensation in the cleansing process, because the reaction is exothermic. The bubbling action is due to the rapid decomposition of the hydrogen peroxide,

$$2\ H_2O_2\ (aq) \rightarrow 2\ H_2O\ (l) + O_2\ (g) \tag{1}$$

This decomposition reaction is the one that you will be examining today. The isolated decomposition of H_2O_2 occurs very slowly at room temperature. However, the reaction can occur very quickly in the presence of a suitable catalyst. A catalyst is a substance that can alter the rate of a chemical reaction without itself being permanently changed.

To better understand how a catalyst may work, let's look at another system. The decomposition of hydrogen peroxide can also be catalyzed by iron(II) ion. In acid solution, iron(II) ion is oxidized in the first step of a two-step reaction mechanism,

$$2\ Fe^{2+}\ (aq) + 2\ H^+\ (aq) + H_2O_2\ (aq) \rightarrow 2\ H_2O\ (l) + 2\ Fe^{3+}\ (aq) \tag{2}$$

This begins a cycle in which iron(II) ion is initially oxidized to iron(III); in the second step of this reaction (equation 3), the iron(III) ion is reduced back to iron(II) ion. Thus, the initial iron(II) ion is regenerated, and the reaction proceeds quickly in this cyclic manner.

$$2\ Fe^{3+}\ (aq) + H_2O_2\ (aq) \rightarrow 2\ H^+\ (aq) + O_2\ (g) + 2\ Fe^{2+}\ (aq) \tag{3}$$

The overall reaction is the sum of these two steps and results in equation (1), the simple decomposition of hydrogen peroxide.

$$2\,Fe^{2+}\,(aq) + 2\,H^+\,(aq) + H_2O_2\,(aq) \rightarrow 2\,H_2O\,(l) + 2\,Fe^{3+}\,(aq) \qquad (2)$$

$$2\,Fe^{3+}\,(aq) + H_2O_2\,(aq) \rightarrow 2\,H^+\,(aq) + O_2\,(g) + 2\,Fe^{2+}\,(aq) \quad (3)$$

$$2\,H_2O_2\,(aq) \rightarrow 2\,H_2O\,(l) + O_2\,(g) \qquad (1)$$

To catalyze the decomposition of H_2O_2 for your experiment, you will use an enzyme called catalase that is derived from bovine (cattle) liver. Catalase is present in many different organisms, and it functions to protect the organism from the destructive—even toxic—reaction between peroxide and animal tissue. The human variant of catalase is the substance responsible for the bubbling that occurs when hydrogen peroxide is poured on an open wound.

In the first part of this experiment, you will measure the heat liberated by the decomposition of several different concentrations of hydrogen peroxide solution. The concentrations will be percent (by mass) solutions. Because hydrogen peroxide is the active ingredient in this reaction, the heat liberated by the reaction is directly proportional to the amount of H_2O_2 present in the sample.

In the second part of this experiment, you will devise and test a procedure that will measure the extent to which you can decrease the heat transferred from the H_2O_2 decomposition reaction by using the substances you studied in the skill-building lab. While it is generally acknowledged that nothing could have survived the 800 °C temperatures that led to the collapse of the World Trade Center towers, the fireproofed columns and floors allowed them to remain standing for almost two hours after the initial impact and explosion. During this time, thousands of people were evacuated from the buildings. The heat capacity of the fireproofing materials acted as a heat sink and delayed the rapid melting of the buildings' structural supports.

Can you create a similar heat sink that will control the temperature change from the heat given off by the H_2O_2 reaction? You won't have time to "build a room" to surround your reaction, but you could measure the heat-storing capabilities of a substance by simply adding it to the reaction mixture. Suppose you add a material that is able to absorb and store some of the evolved heat. Would this lower the final temperature you measure? How does heat capacity enter into your plan? You measured the heat capacity for several kinds of materials in the skill-building lab. Which of these materials will you use? How much of it will you use?

Analysis of Data

The analysis of the data for this experiment is similar to that in the skill-building lab. Once again, the system is adiabatic. But now we can write a heat change associated with the reaction, q_{rxn}:

$$q_{univ} = 0 = q_{rxn} + q_{cal} + q_{H_2O}$$

Here q_{rxn} is the heat liberated by the decomposition of hydrogen peroxide, q_{cal} is the heat absorbed by the calorimeter, and q_{H_2O} is the heat absorbed by the water in the solution. Assume this is the mass of the entire peroxide solution. You will have experimental data with which to calculate q_{cal} and q_{H_2O}. Assume the specific heat of water in the H_2O_2 solution is the same as that of pure water, $4.184\ J\ g^{-1}\ K^{-1}$.

All of the heat generated or absorbed by the reaction must be offset by heat absorption or liberation by the calorimeter and its contents. Therefore, solving for q_{rxn} we get

$$q_{rxn} = -(q_{cal} + q_{H_2O})$$
$$q_{rxn} = -(C\,\Delta t)_{cal} - (c_s m\,\Delta t)_{H_2O}$$

You know the mass of the hydrogen peroxide solution. The initial and final temperatures of the calorimeter and of the hydrogen peroxide solution are the same; the two are in contact at the start of the run. Remember that the heat capacity of the calorimeter is C_{cal}, not q_{cal}. Once you know q_{rxn}, you can calculate the molar heat of reaction, ΔH_{rxn}:

$$\Delta H_{rxn} = q_{rxn}/n$$

where n is the number of moles of H_2O_2 actually used in each experiment.

CAUTION: Although the solutions of hydrogen peroxide you will be using are very dilute, you should always observe proper laboratory safety precautions when handling chemicals of any kind. Avoid chemical contact with your skin and especially with your eyes. Safety goggles should be worn at all times in the lab.

Procedure

Part I: Formation of Groups

For group discussion and data analysis purposes, this experiment is best done in groups of three. Each individual in the group is responsible for building and calibrating his or her own calorimeter and for measuring one of the three concentrations of H_2O_2 available in the lab. There will be two samples of each concentration to measure. At the completion of the experiment, the group will pool the data for the different concentrations of hydrogen peroxide and compare the heats of reaction obtained.

  When you and your group are confident in your ability to measure the heat of the H_2O_2 reaction, you will plan and write up a brief procedure to decrease the heat given off by the H_2O_2 reaction. Include any precautions you will follow that will ensure the quality of your measurements. When your instructor has approved your plan, you will test it by measuring the heat of the H_2O_2 reaction again.

Part II: Construction of a Calorimeter (Individual Work)

Each member of the group should construct and measure the heat capacity of a calorimeter. It is necessary to determine C_{cal} for each new calorimeter you construct and use. (If necessary, refer to the directions given in the skill-building lab.)

CAUTION: If your calorimeter is made of a beaker inside of a metal can, be sure to put the H_2O_2 solutions into the beaker and *not* into the metal can.

Part III: Heat of Reaction of Hydrogen Peroxide (Individual Work)

Each member of the group should work with a different concentration of hydrogen peroxide and should do two trials—one with approximately 50 g of solution and one with 75 g of solution. The concentration of the peroxide solutions will be expressed as mass percentage. Obtain the mass of the H_2O_2 solution. Make all mass measurements to ±0.01 g.

  Assemble the calorimeter with the H_2O_2 solution, cover, and insert the thermometer. Wait until the temperature stabilizes. Record the temperature to ±0.2 °C.

  When you are ready, add three to four drops of the catalase solution. Immediately replace the cover and thermometer in the calorimeter, and *swirl gently to mix well*.

CAUTION: Do not use the thermometer to stir the solution as it is fragile and may break. Instead, rotate the calorimeter in small circular motions so that the solution inside is gently mixed.

  Watch the thermometer until the temperature reaches a maximum. The reaction occurs rapidly (in 2–4 min) if the catalase is freshly prepared. Record the maximum temperature attained. Repeat the reaction using the same calorimeter. This time, use a different mass of the peroxide solution.

Part IV: Devising and Testing a Plan to Control Temperature (Group Work)

Now that you have some experience measuring the heat of reaction for the decomposition of hydrogen peroxide, you and your group are ready to devise and test a plan that will lower the measured heat. Clearly and neatly write the steps you will follow for this part of the experiment. Be sure to specify which concentration of H_2O_2 you will test; include appropriate masses and heat capacities of the substances you use.

When your plan is complete, show it to your instructor for approval. Then measure the heat of reaction for the hydrogen peroxide decomposition again, this time following the procedure you have agreed upon.

Report

Calculations

1. Determine C_{cal}, the heat capacity, for your calorimeter. Use an averaged value of C_{cal} for the rest of the calculations.
2. Calculate q_{rxn} for the concentration of peroxide you used. Do a calculation for both sample sizes. Express your answers in kJ.
3. Calculate ΔH_{rxn} for each solution you measured. Express your answers in kJ/mol.
4. Use standard heat of formation values (ΔH_f°) obtained from tables to calculate the theoretical value for the heat of reaction for $2\ H_2O_2 \rightarrow 2\ H_2O + O_2$. Express your answer in kJ/mol.
5. Compare your experimental value of ΔH_{rxn} with the theoretical value just calculated to determine the percentage error in your measurements.
6. For reactions in which the pressure or volume do not change, the First Law reduces to $\Delta E = \Delta H_p$, but for reactions that generate a gas, this is not true. Then ΔE is calculated using the full equation, $\Delta E = \Delta H - P\,\Delta V$. Calculate ΔE for your reaction when T has a value of 313 K. Remember to make the appropriate substitution for PV.

Group Work

A. Analysis of Data for ΔH_{rxn} (Part III)

Compile group results and discuss the following questions. Include the table and your responses to the questions in your individual report.

1. Prepare a table to summarize the data obtained by your group. Each group member should contribute the following information for this table: percent by mass concentration of the samples you measured, the actual mass of H_2O_2 used in your samples, q_{rxn} for each sample, ΔH_{rxn} for these samples, and the percentage error. (See calculations 2, 3, and 5.)
2. Discuss the contents of these tables as a group, and formulate a group interpretation based on the values of q_{rxn} and of ΔH_{rxn} obtained.
3. Discuss the possible sources of error in this experiment. Make a list of three of these errors and include a prediction of how each error might affect the final result. (High result? Low result? No effect?)
4. Each member of the group should use uncertainty analysis to calculate the uncertainty in q_{rxn} due to the uncertainty in q_{H_2O}. (This is only part of the calculation you would have to do to find the total uncertainty in q_{rxn}, but it will serve as an example of the analysis.)

B. Analysis of Your Plan for Measuring ΔH_{rxn} (Part IV)

1. Briefly summarize the plan your group devised. Include all pertinent data and calculations of ΔH_{rxn}.

2. Discuss the strengths and weaknesses of your plan. Did your plan succeed in lowering Δt? Are there any modifications you can suggest for this procedure?

3. What did you think was the most difficult part of this procedure? Why?

Experiment Group K

Ecological Element Cycles: Judge Levels of Environmental Disturbance

The application of nitrogen is common on many farms.
(Larsh Bristol/Visuals Unlimited.)

Purpose

In this experiment group, you will utilize two different analytical methods—instrumental and titrimetric—to perform elemental analyses of two different ecologically important elements, phosphorus and nitrogen. Determining elemental composition is not just an academic activity. We need to keep track of the amount of elements if we are to study the movement of elements through the ecosystem.

- *Instrumental techniques.* An instrument is used to measure the quantity of analyte (the substance being determined) present in the sample. The most common instrumental techniques utilize spectrophotometry (such as UV-visible, infrared, and atomic absorption) or chromatography (such as gas chromatography [GC] and high-performance liquid chromatography [HPLC]).

- *Titrimetric methods.* These involve measuring the volume of a solution of known concentration needed to react with the analyte or some derivative of the analyte.

The method that a scientist uses depends on several factors. The most important are accuracy, cost, and convenience. Instrumental methods may be more expensive in the short run, but they may be much easier and more reliable. On the other hand, so-called wet methods, such as titrimetric methods, are often less expensive and easier to do in the field. The nature of the analyte also affects this decision. In spectrophotometry the analyte must be completely converted into some species that absorbs light without interference. In a titration, one needs a method to measure the change in solution composition that signals the end point.

Schedule of the Labs

EXPERIMENT 1: Skill-Building Lab: Analysis of Phosphorus in Water
- Prepare a calibration curve using standard solutions (group work).
- Determine the concentration of phosphorus in a prepared unknown (group work).
- Determine the concentration of phosphorus in a fresh water sample (individual work).
- Determine the concentration of phosphorus in a fresh water sample after it has been disturbed (individual work).

- Set up the soil analysis experiment to be completed the second week of the experiment group (individual work).

EXPERIMENT 2: Application Lab: Nitrogen and Mud

- Determine the percentage by mass of mineral nitrogen in disturbed and undisturbed soil samples (individual work).
- Determine the mineral nitrogen content of a standard solution (individual work).
- Dry a soil sample (individual work).
- Organize and analyze group and class results (group work).

Scenario   Your family decides to build a cabin—a place to go in the summer, up north, on a lake in the woods. "Will building the cabin disturb the environment?" your mom wonders. Your dad, remembering the cost of tuition, says "From what she learned in college, she should be able to tell us."

You're on the spot, and you remember some techniques you learned in introductory chemistry that would be helpful. You explain that it is important to measure the response of two key elements that all organisms need for growth and reproduction: nitrogen and phosphorus. Nitrogen is needed to construct amino acids and the enzyme that plants use to fix carbon; phosphorus is needed for important molecules such as DNA and ATP.

"What exactly do we measure?" your mom asks. You explain by drawing the cycles of nitrogen and phosphorus for the site in the woods and the lake where the cabin is to be built (Figure K-1). You point out the places in the cycles where nitrogen is in inorganic form—NH_4^+, NO_3^-, and NO_2^-—and where phosphorus is in inorganic form—PO_4^{3-}. There are two reasons to focus on these inorganic compounds, you explain. First, inorganic compounds will be relatively easy to measure chemically

Figure K-1 Inorganic nitrogen and phosphorus cycle

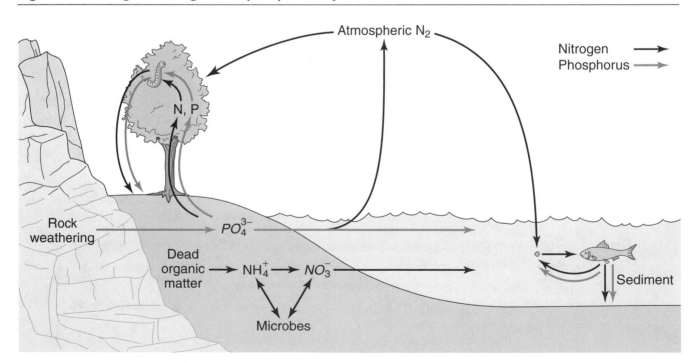

because they are not parts of complex organic molecules that would be difficult to digest before measurement. Second, inorganic nitrogen and phosphorus are the immediately available "food" of bacteria, fungi, and roots, so they are important.

"How do we go about taking the measurements?" asks your dad. You remember a diffusion method for measuring NH_4^+ and NO_3^- (known as mineral nitrogen) in small water or oil samples. You also remember a method for sampling dissolved phosphate using a colorimeter. So that is what you suggest.

What you do *not* remember is what indicates environmental disturbance, and that is what your mom is really concerned about. So you admit that you aren't sure whether the quantities of NH_4^+, NO_3^-, and PO_4^{3-} in the lake should be higher or lower after the cabin is built. So you decide to compare the soil and lake water before and after your family summer home is built.

EXPERIMENT 1
Skill-Building Lab: Analysis of Phosphorus in Water

Pre-Laboratory Assignment   **Due Before Lab Begins**

NAME: _____

Complete these exercises after reading the experiment but before coming to the laboratory to do it.

1. What is the purpose of a blank in a spectrophotometric analysis? What will serve as the blank in this experiment?

2. What will be the concentration of phosphorus in a standard solution prepared by diluting 10.00 mL of 2.5 mg/L of phosphorus solution to 25.00 mL?

3. When preparing the cuvette to analyze either standard or sample, 86.2% of the total volume in the cuvette should be the phosphorus solution, and 13.8% of the total volume should be the mixed reagent. If your cuvette contains a maximum volume of 8.00 mL, how many milliliters of phosphorus solution and how many milliliters of mixed reagent will you add to the cuvette?

4. Experimentally, what should you do if your unknown's concentration of phosphorus is outside the range of concentrations (either too high or too low) of the standards?

5. A stock phosphorus solution is prepared by dissolving 219.5 mg of anhydrous KH_2PO_4 in water and diluting to 1.00 L. The bottle is labeled as 50.0 µg P/mL. Show that the label on the bottle is correct.

6. Consult the MSDS (material safety data sheet) for the chemicals used in the mixed reagent and note any special precautions you should take while handling these solutions.

EXPERIMENT 1
Skill-Building Lab: Analysis of Phosphorus in Water*

Background

Phosphorus is an essential element for plant growth. While the phosphate ion, PO_4^{3-}, is fairly common in soil, mostly as an insoluble salt, its concentration in surface fresh water (lakes, streams, and rivers) is often less than that of many other dissolved nutrients. Thus, phosphorus, present as the inorganic phosphate ion, often serves as the limiting nutrient necessary for algal growth.

When phosphate levels are high, aquatic plant growth increases and algal blooms may occur. As the algae die and decompose via oxidation, the dissolved oxygen in the water decreases and fish life is threatened. The lake water can also become green, slimy, and foul as dead fish and plants rot on the shore. This was the major problem with Lake Erie when it was considered a "dead" lake. Detergents were once the major source of phosphorus in fresh surface water, but environmental concerns over too much phosphorus in bodies of water led to the development of phosphate-free detergents. Other sources of phosphorus in water systems include septic tanks, runoff from feedlots, agricultural runoff, and waste water treatment plants.

The concentration of phosphorus in water can be measured spectrophotometrically. Ammonium molybdate reacts with the phosphate ion in an acidic environment to produce phosphomolybdic acid. The phosphomolybdic acid is then reduced by ascorbic acid to molybdenum blue. The absorption of this intensely blue complex is measured and, through the use of a calibration curve, the concentration of phosphorus is determined.

Procedure

Important: Your instructor may ask you to bring to class a sample from a body of fresh water to analyze for phosphorus.

Part I: Formation of Groups

You will be working in a group of three people. Everyone is expected to participate in the experiment, to record data into his or her own notebook, and to complete his or her own lab report. You will prepare the calibration curve together and then you will individually analyze a fresh water sample for its phosphorus concentration.

Part II: Preparation of the Calibration Curve (Group Work)

Good analytical procedure requires that all samples be treated in the same manner. This involves the following four steps:

1. Prepare solutions in volumetric flasks containing known amounts of the species you are studying.
2. Transfer the same amount of each solution to a cuvette.
3. Add the same amount of the derivatizing agent to the cuvette. Cap and mix well.
4. Measure the absorption spectrum of the solution in the cuvette.

---

*The method for determining the phosphorus concentration in water is a standard analysis described in Eaton, A. D., Clesceri, L. S., and Greenberg, A. E. (eds.), *Standard Methods for the Examination of Water and Wastewater*, 19th ed., 1985, pp. 4-113 and 4-114.

Therefore, your first step is to prepare your set of phosphorus solutions. Prepare at least six different solutions beginning with the stock phosphorus solution. Use a separate volumetric flask for each solution. These are called standard solutions because their concentrations are known.

Next, you need to prepare what is referred to as the mixed, or combined, reagent. This is the solution that contains both the ammonium molybdate and ascorbic acid needed to form the intensely colored blue complex. In a 100-mL volumetric flask, add 50 mL of 2.5 M H_2SO_4, 15 mL of the ammonium molybdate solution, and 30 mL of the ascorbic acid solution. Bring the total volume to 100 mL with deionized water. Cap and mix thoroughly. If turbidity forms, shake the solution and let it stand for a few minutes until the turbidity disappears before proceeding.

Before preparing the solutions for which you will measure absorbances, prepare the spectrophotometer. Adjust the filter wheel so that the red filter is in place and set the wavelength for 880 nm. Be sure to adjust the instrument to register both 0 and 100% transmission.

Find out the volume of the cuvette used in your spectrophotometer. Of the total volume in the cuvette, 86.2% will be the phosphorus solution and 13.8% will be the mixed reagent. First add the appropriate volume of each of the standard phosphorus solutions to separate cuvettes, then add one drop of phenolphthalein solution to each. If any of the solutions turn pink or red, add 2.5 M H_2SO_4 dropwise until those solutions are colorless. Finally, add the calculated amount of the combined reagent to each cuvette. Cap the cuvettes and mix thoroughly. After at least 10 min, but no more than 30 min, measure the absorbance of each standard at 880 nm.

Recall that good analytical technique requires that you treat each sample the same way, so be sure to measure the absorbance of each solution after the same amount of time. Plot a calibration curve showing absorbance as a function of phosphorus concentration. Include the best-fit equation and the correlation coefficient. Examine the calibration curve. If any of the points are grossly out of line from the curve or the curve is not close to passing through the origin, decide what steps you should take to prepare a more reliable calibration curve.

Part III: Determining the Phosphorus Concentration in a Prepared Unknown (Group Work)

Prepare the unknown solution the same way you prepared the standards. Be sure to run multiple trials on the unknown (to check for reproducibility). Use your group's calibration curve to determine the phosphorus concentration of the samples. Share your results with your instructor and decide whether the calibration curve is acceptable or whether you need to work to improve it.

Part IV: Determining the Phosphorus Concentration in a Fresh Water Sample (Individual Work)

Once you have an acceptable calibration curve, prepare your fresh water sample the same way you prepared the standards. Be sure to run multiple trials on the sample (to check for reproducibility). Use the final version of your group's calibration curve to determine the phosphorus concentration of the samples.

Part V: Determining the Phosphorus Concentration in a "Disturbed" Fresh Water Sample (Individual Work)

Give your instructor a sample of your fresh water. Your instructor will "disturb" the sample in some way and return it to you. You will then measure the phosphorus concentration of the disturbed sample.

Report

Tabulate the absorbance and phosphorus concentration for each of the analyses you completed on your own sample. Report the source of the water and your average phosphorus concentration to the class. Tabulate class results.

Discuss the reliability of your calibration curve. Use experimental and statistical results to support your discussion. Obtain a copy of the calibration curve from at least two other groups. How do your best-fit equations and correlation coefficients compare? How did you expect them to compare? Explain. What would happen to the value for the concentration of phosphorus in your water sample if you used one of the other calibration curves? Explain. Which calibration curve should be used for you to best analyze the sample? Explain.

For each fresh water sample with a measurable phosphate concentration, offer an explanation as to the source of the phosphate. Some municipalities add phosphate to their drinking water supplies. If your fresh water sample is from a drinking water source, investigate why phosphate is used as an additive. Explain whether or not your results for the phosphate concentration in the drinking water support community policy on phosphates as an additive to drinking water.

Explain the impact of the disturbance on the phosphorus concentration. Identify possible sources of the disturbance.

Although you diluted each of your standards when you prepared the cuvette, you did not have to account for this dilution when preparing your calibration curve—as long as you treated your samples in exactly the same manner as the standards. Explain why this is the case.

Preparation for the Application Lab

In the application lab, you will use acid–base titration to determine the amount of mineral nitrogen in soil. Each student must set up four samples for analysis. Follow this procedure.

Most NH_4^+ ions in soils are attached to negatively charged sites on the edges of clay or organic matter particles. Potassium ions, K^+, bind to these sites more strongly than ammonium ions. Therefore, the ammonium ions can be released into solution by treating the soil sample with a KCl solution. Nitrate ions present in the soil are reduced to ammonium ions by the addition of a reducing agent known as Devarda's alloy. This alloy is a mixture of metals consisting of 50 parts Cu, 45 parts Al, and 5 parts Zn. Then, by making the solution basic, the NH_4^+ ions are volatilized from the solution as ammonia, NH_3.

The ammonia volatilizes and diffuses slowly into the air. A boric acid (H_3BO_3) solution, kept in proximity to but not in direct contact with the soil sample, will absorb the ammonia. An indicator added to the boric acid solution changes color as the NH_3 is absorbed. Prepare four small beakers or specimen cups with about 15 mL of the boric acid–indicator solution. Also, obtain four screw-top jars. Make sure they are clean and dry.

In the first jar, place 2 to 3 g of fresh undisturbed soil. Record the mass of the soil used to ±0.001 g. In the second jar, place 2 to 3 g of fresh disturbed soil. Record the mass of the soil used to ±0.001 g. To *each* of your soil samples, add about 20 mL of 2 M KCl solution, sprinkle in about 0.2 g of Devarda's alloy, and add approximately 0.2 g of solid MgO (to raise the pH). Neither the Devarda's alloy nor the MgO need to be weighed accurately. Swirl to mix.

Place one of the boric acid–indicator samples in each of the jars, then seal them.

A control must also be set up in this experiment. The control jar is identical to the sample jars *with the exception of the soil.*

It is important to check your experimental design for complete recovery of mineral nitrogen by analyzing a standard solution (a solution in which the quantity of mineral nitrogen is known). Set up another jar in which you replace the soil with 2.00 mL (measured exactly) of the standard solution. Swirl to mix.

The systems must sit undisturbed for a week. Place all of your jars in a location specified by your instructor.

EXPERIMENT 2
Application Lab: Nitrogen and Mud

Pre-Laboratory Assignment   **Due Before Lab Begins**

NAME: _____

Complete these exercises after reading the experiment but before coming to the laboratory to do it.

1. You will titrate a boric acid solution that has absorbed the ammonia from a soil sample. It is not necessary to know the exact volume or concentration of the boric acid. Explain why not. [*Hint*: Combine equations (1) and (2) in the laboratory procedure.]

2. To prepare for this week's lab, you added a KCl solution, Devarda's alloy, and MgO to your soil sample in order to analyze for mineral nitrogen. Sketch a jar containing soil and indicate how the nitrogen is initially present in the soil. Then draw three more jars showing the impact of each of the three substances— KCl, Devarda's alloy, and MgO—on the soil sample.

3. (a) Calculate the number of moles of ammonia absorbed by a boric acid solution if 3.25 mL of 0.0025 M H_2SO_4 were needed to titrate the solution to the end point. (Assume both hydrogens of the sulfuric acid react.)

   (b) From the number of moles of ammonia calculated in part (a), determine the number of micrograms of nitrogen per gram of dry soil. Assume you have 2.753 g of soil and all of the ammonia diffused from the soil.

4. If some of the acid solution drips down the side of the stock bottle as you fill your own beaker, what should you do?

EXPERIMENT 2
Application Lab: Nitrogen and Mud*

Background

Nitrogen and phosphorus are the two nutrients that most often limit the growth of plants. For this reason, much effort has been put into finding ways to measure nitrogen and phosphorus in soil and water. Nitrogen is present in soil and water in both organic compounds (primarily in amino acids) and inorganic compounds (as ammonium, NH_4^+, and nitrate, NO_3^-, ions). The nitrogen present in inorganic compounds is known as mineral nitrogen. Microorganisms, in a process known as nitrification, rapidly catalyze the oxidation of ammonium ions first to nitrite ions (NO_2^-) and then to nitrate ions (with some N_2O as a byproduct). These microorganisms compete for the ammonium ions with roots in soil and with algae in water.

You will be determining the percentage by mass of mineral nitrogen in two samples of soil, one that has been disturbed and one that is undisturbed, through an indirect titrimetric method. Thus, you will measure the amount of nitrogen available for a plant root, bacterium, or fungi to take up and you can determine the impact of the disturbance on the mineral nitrogen content. To analyze for mineral nitrogen, you need to separate the mineral nitrogen from the rest of the sample. The method you will use to accomplish this separation involves the following steps:

1. Displacement of NH_4^+ and NO_3^- ions from the soil by KCl
2. Conversion of NO_3^- to NH_4^+ with alloy
3. Volatilization of the NH_4^+ as NH_3
4. Absorption of the resulting NH_3 vapor by a boric acid solution
5. Titration of the resulting solution with an acid

> **NOTE:** This experiment must be set up one week prior to the analysis to allow time for the separation of the mineral nitrogen. You also need to dry a separate portion of your original fresh soil sample in a drying oven so that you can report your results per gram of dry soil. Weigh a 2- to 3-g sample of wet soil, record the mass to 0.001 g, and place it in the drying oven.

CAUTION: You will be titrating with a dilute solution of a strong acid during this experiment. Clean up small spills (several drops) with excess water. Neutralize large spills (several milliliters) with sodium hydrogen carbonate or as indicated by your instructor.

Procedure

In the samples you set up last week, the mineral nitrogen, present as ammonium and nitrate ions, has been volatilized into the atmosphere of the jar as ammonia gas. The boric acid solution in the jar absorbs this ammonia. An indicator added to the boric acid solution changes color as the NH_3 is absorbed.

---

*The method for determining the mineral nitrogen content in soil utilized in this experiment is described in Mulvaney, R. L., "Nitrogen—Inorganic Forms," pp. 1123–1184 in D. L. Sparks et al., editors, *Methods of Soil Analysis*. Part 3, Soil Science Society of American Book Series 5. ASA and SSSA: Madison, WI, 1996.

$$H_3BO_3 \; (aq) + NH_3 \; (g) \rightarrow H_2BO_3^- \; (aq) + NH_4^+ \; (aq) \tag{1}$$

The indicator is a mixture of bromocresol green and methyl red. Record the initial color of the mixed indicator in the boric acid solution. The indicator will turn green as the solution absorbs ammonia.

The resulting $H_2BO_3^-$ can then be titrated back to the original color of the indicator by an acid that provides hydrogen ions.

$$H^+ \; (aq) + H_2BO_3^- \; (aq) \rightarrow H_3BO_3 \; (aq) \tag{2}$$

The amount of acid needed to reach the end point is used to determine the amount of nitrogen present in the original sample. This is known as an indirect method because the ammonia, the compound of interest, is not directly titrated.

To complete the analysis for nitrogen, titrate the boric acid solution that absorbed the ammonia from your samples, the control, and the standard.

A pair of students should practice this titration together, but each student will be responsible for analysis of the four samples prepared last week.

To practice the titration, place about 20 mL of the boric acid solution in a beaker. Use a volumetric pipet to add a known amount of standardized ammonia solution to the beaker, and swirl gently to mix. Note the color change. Titrate this solution with standardized acid to the end point. This should be done at least twice to be sure that the end point is reproducible. Then you are ready to titrate each of the four experiment setups.

> **NOTE:** You dried a separate sample of the same soil you are analyzing for mineral nitrogen. You needed to do so in order to determine the moisture content of the soil. Your results of the analysis for mineral nitrogen are to be reported in micrograms of nitrogen per gram of dry soil. The sample you will analyze for mineral nitrogen is not dry, but if you know the percentage, by mass, of water in an equivalent soil sample, you can calculate the dry mass of the sample used in the analysis.

Calculations

Calculate the percentage (by mass) of water in your soil sample. Calculate the dry mass of the soil sample used in your analysis for mineral nitrogen.

Determine the number of moles of acid used in your titration of the sample and the control. Use this value to calculate the number of moles of ammonia absorbed by the boric acid. Finally, calculate the number of micrograms of mineral nitrogen present in your soil sample. Be sure to factor in the control. Report your results as micrograms of nitrogen per gram of dry soil.

Calculate the number of moles of acid used in your titration of the standard. Use this value to determine the number of moles of ammonia absorbed by the boric acid. Finally, calculate the experimental number of micrograms of nitrogen per milliliter of solution in the standard. Obtain the expected value from your laboratory instructor and calculate your percentage recovery.

Report

Tabulate all results and show sample calculations.

Describe the experimental design used in this experiment and comment on the validity of the design to accomplish the goal of the experiment. Explain the role of the control in this experiment. Explain how you were able to calculate the mass of dry soil used in the analyses.

Share the results of your soil analysis and your percentage recovery of the nitrogen in the standard with the entire class. Draw conclusions regarding the effectiveness of the method used in this experiment for analysis of mineral nitrogen. Also discuss the impact of the disturbance on the mineral nitrogen concentration in the soil. Support your conclusions with data.

If your percentage recovery of mineral nitrogen in the standard solution is not 100% ± 2%, explain what factors may have contributed to the recovery error.

Questions to Answer in Your Report

What about the original question you couldn't answer in the Scenario? If you build a cabin in the forest, will the disturbance of construction and regular use raise or lower the phosphate, ammonium, and nitrate concentrations in water and soil? Answer the scenario question based upon your responses to the following questions. Include your answers to all questions in your final report.

1. Does disturbance of soil increase or decrease the concentration of ammonium or nitrate? Use your experimental results to support your answer.

2. If a soil is disturbed for years, what is likely to be the effect on soil fertility? Explain.

3. Atmospheric deposition of important ions is recorded by the National Atmospheric Deposition Program (NADP). Look for their Web site at *http://nadp.sws.uiuc.edu*. Using isopleth maps of ammonium and nitrate at the NADP Web site, look at maps of ammonium and nitrate deposition. Where are the centers of ammonium deposition? Where are the centers of nitrate deposition? Can you infer sources of ammonium and nitrate in our atmosphere from the maps? Explain.

Experiment Group L

Equilibrium in Aqueous Solutions

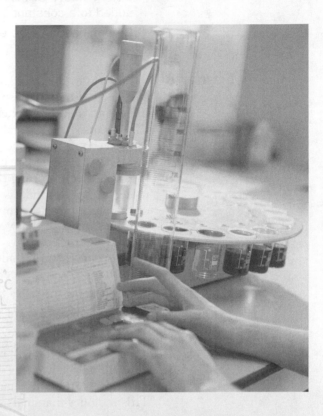

Biochemists use instrumentation to determine the chemical interactions important in life. (DigitalVision.)

Purpose Many very important chemical reactions do not proceed to the completion suggested by the stoichiometry of the reaction. As a result, reaction mixtures become mixes of starting materials and products. This is a situation where *chemical equilibrium* applies.

Equilibrium is far more important than the label "incomplete reaction" implies. A system at equilibrium can respond to small and large stresses and can return to almost the original conditions. This flexibility is important in keeping systems stable, a topic that is explored in detail in Experiment Group G, "Buffers and Life," which you might do this term.

Equilibrium is also involved in determining just how much of a chemical species will be present under particular conditions. In this case, an equilibrium can be manipulated to soak up or to selectively release a substance from its environment. In other cases, an equilibrium can determine how much reactant and how much product are present, depending on other species present. Therefore, it is important that chemists be able to determine the way an equilibrium reaction will behave. This means obtaining data that let us find the equilibrium constant for a reaction.

Experiment 1: Skill-Building Lab: K_a of an Indicator

Schedule of the Labs
- Determination of the pH transition range for a series of indicators (group work).
- Microscale determination of the approximate range of pK_a of the indicator (individual work).
- Spectrophotometric determination of the pK_a of the indicator (individual work).

Experiment 2: Application Lab: Equilibrium of Protein–Ligand Binding

- Determination of the appropriate concentrations of substrate and protein for study (group work).

- Determination of the absorbance change when different amounts of protein are added to a constant amount of substrate (individual work).

- Determination of the effect of an interfering molecule on the equilibrium (individual work).

Scenario   When a nutrient, drug, or other small molecule is absorbed by an organism or by a cell, it is usually transported to its cellular destination by a larger molecule. An important area of biochemistry and medicinal chemistry is the determination of how this occurs. Because transport requires uptake in one area of the body (such as the digestive tract) and release in another (such as the brain), the transport mechanism must be reversible. In chemistry, such reversible processes are usually equilibrium reactions. Determining the equilibrium involved in the uptake and release of a small molecule by a protein is essential. This kind of equilibrium is often called an *affinity* of the protein for the small molecule.

Let's imagine that you are a biochemist studying the way that a new drug can be used in a clinical setting. You know that the drug works in studies of isolated cells, but you also need to find out if the body's transport mechanism can function to deliver the drug to the target cells. Among the questions you must ask are:

- If you know the likely transport protein, will it bind the drug?

- What is the affinity of the drug for the protein?

- Can other small molecules interfere with the affinity?

In this lab, you will use the tools of spectrophotometry to answer these questions. You will not be working with a real pharmaceutical, but the protein is one that is used in many vertebrates that have to transport molecules through a circulatory system. This includes mammals such as the cattle that are the source of the protein you will use today.

EXPERIMENT 1
Skill-Building Lab: K_a of an Indicator

Pre-Laboratory Assignment   **Due Before Lab Begins**

NAME: _____

Complete these exercises after reading the experiment but before coming to the laboratory to do it.

The data graphed in Figure L-1 were collected for the indicator phenolphthalein. This indicator is colorless at intermediate pH values and becomes magenta at high pH. The absorbance spectra of a series of solutions of phenolphthalein at different pH values is shown in the figure.

Figure L-1 Changes in the absorbance spectrum of phenolphthalein at different pH values

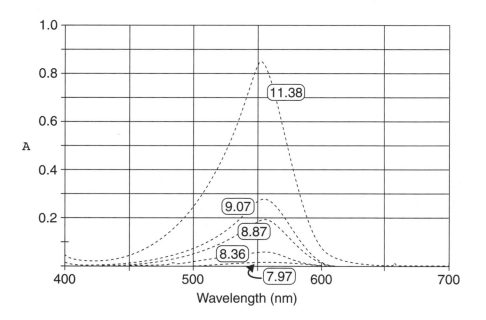

1. For the four intermediate pH's in the graph, determine the value of the fractional ionization α of HA at the pH's listed. Use these values to determine four independent values for K_a. Present your data by filling in the following table.

pH	A^{550}	α	K_a
7.97			
8.36			
8.87			
9.07			
11.38 ("pure base")			

2. Prepare a graph of the absorbance versus pH of phenolphthalein at 550 nm in the pH range 7 to 8.5. Indicate the approximate point where the absorbance is half of the value at pH = 11.38.

3. What particular safety measures must you take in this laboratory to avoid a false sense of security?

EXPERIMENT 1
Skill-Building Lab: K_a of an Indicator

Background

Many important reactions do not proceed to completion when reactants are mixed. Instead, a ratio of species exists in a chemical equilibrium. In these cases, spectrophotometry can be very important in determining the relative amounts of different substances. This experiment uses the experimental strategies of tissue culture plates and spectrophotometry to deduce the ratio of different forms of a compound that is affected by the acid and base properties of a solution.

An important equation describing the action of an acid in water relates the concentrations of hydrogen ion, H^+, an acid, HA, and the acid's conjugate base, A^-, to K_a, a constant:

$$K_a = \frac{[H^+][A^-]}{[HA]}$$

This relationship holds at all times for a system at equilibrium. The concentration of H^+ may be established through the reaction of the acid HA or the base A^- with water, or the concentration of H^+ could be set by some chemical system so dominant that the concentrations of A^- and HA are forced to follow along. In fact, in research and technological applications it is often more important to ask the question "How does $[H^+]$ influence this chemical substance?" instead of "How does this chemical substance influence $[H^+]$?"

Acid–base indicators are prominent examples of substances that have little influence on pH but whose properties, in their case color, are influenced by $[H^+]$. Their color makes them good candidates for the determination of their equilibrium properties by using visible light and a spectrophotometer.

Indicators

Indicators are weak acids or weak bases that undergo one or more ionization steps in the range between pH = 0 and pH = 14. All weak acids have a range of pH's for which there is a significant amount of the substance in the acid form (HInd) and a significant amount in the base form (Ind$^-$). An indicator will undergo a sensible color change in this range. In the middle of the transition, the mixture of the ionized and un-ionized forms of the indicator will give a mixed color.*

The color of an indicator can suggest how much of it is in the base form and how much is in the acid form. For example, the indicator bromocresol green is yellow in acid solution and bright blue in basic solution. However, between a pH of about 3.8 and about 5.2, both the acid form and the base form are present. Therefore, the eye perceives the color that is a mixture of blue and yellow, which is green.

The equilibrium for an indicator in a solution of controlled pH can be described by the chemical equilibrium equation

$$\text{HInd } (aq) \rightleftarrows \text{Ind}^- (aq) + H^+ (aq)$$

---

*The sharp, dramatic color change we need when an indicator is used in a stoichiometric titration is caused by the sudden change in pH from a value well below to a value well above the transition range. This can happen with a fraction of a drop of titrant. The pH changes in this experiment are much more gradual to permit study of the intermediate colors.

This suggests a relationship among the concentrations of Ind⁻, HInd, and H⁺:

$$K_a = \frac{[\text{Ind}^-]}{[\text{HInd}]} [\text{H}^+]$$

It is easy to determine the concentration of the hydrogen ion, $[\text{H}^+]$, by using a pH meter. However, we also need information about $[\text{Ind}^-]$ and $[\text{HInd}]$. The "trick" of this lab lies in the fact that we don't need to know what these are. We just need to know the *ratio* of $[\text{Ind}^-]$ to $[\text{HInd}]$. This is quite easy to do, if all solutions that we study have the *same* total concentration of indicator: [total indicator] = $[\text{Ind}^-]$ + $[\text{HInd}]$.

Let's say a fraction α of the indicator is in the base form. Then the concentration of the base form will be $[\text{Ind}^-] = \alpha$[total indicator]. The fraction of the indicator in the acid form will be $1 - \alpha$, so $[\text{HInd}] = (1 - \alpha)$[total indicator].

That means that, in the equilibrium expression given,

$$K_a = \frac{\alpha[\text{total indicator}]}{(1 - \alpha)[\text{total indicator}]} [\text{H}^+] = \frac{\alpha}{1 - \alpha} [\text{H}^+]$$

Absorbance Spectra and Fractional Ionization

Figure L-2 shows the changes in the absorbance spectrum for the indicator thymol blue at relevant pH values, and Figure L-3 shows the acid–base equilibrium for thymol blue. All the solutions have identical concentrations of the indicator. Below pH = 7.44, we see that very little visible light is absorbed by the indicator at 590 nm (here, $\alpha = 0$). At pH = 10.40, there is a large amount of light absorbed for this wavelength (here, $\alpha = 1$). In between, the amount of light that is absorbed changes because different amounts of the indicator are in the form that absorbs light at 590 nm. Assuming that thymol blue obeys Beer's law, then the light absorbed is directly proportional to the concentration of the thymol blue in the *base* form.

High pH: $\quad A^{590} = A^{590}_{\text{Ind}^-, \text{pure}} \qquad \alpha = 1$

Low pH: $\quad A^{590} = A^{590}_{\text{HInd, pure}} \qquad \alpha = 0$

Figure L-2

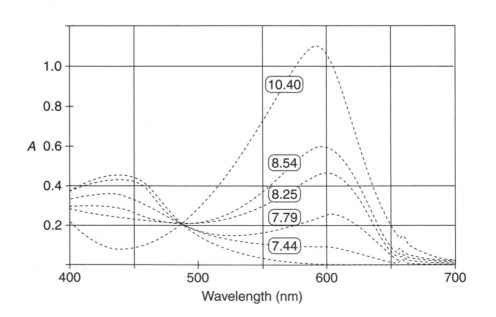

Figure L-3 Acid–base equilibrium for thymol blue

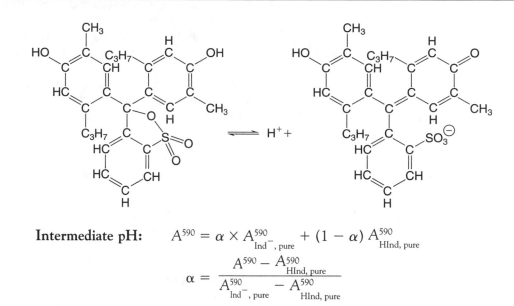

Intermediate pH:

$$A^{590} = \alpha \times A^{590}_{Ind^-, pure} + (1 - \alpha) A^{590}_{HInd, pure}$$

$$\alpha = \frac{A^{590} - A^{590}_{HInd, pure}}{A^{590}_{Ind^-, pure} - A^{590}_{HInd, pure}}$$

So, we now have a way to get the fraction ionized, α. We just compare the absorbance at some intermediate pH with that at high pH and low pH.*

Let's look, for example, at the thymol blue data. We can determine that the absorbance at pH = 10.40 is 1.12. We also see that the absorbance of the acid form is about 0.02 at pH < 6.00. We can compare the absorbance at other values to get the value for α at those pH values:

pH	A^{590}	α
7.44	0.10	0.07
7.79	0.23	0.19
8.25	0.43	0.37
8.54	0.59	0.52

This can now be used to get values for K_a at different pH values:

pH	α	$[H^+]$	K_a
7.44	0.07	3.6×10^{-8}	2.7×10^{-9}
7.79	0.19	1.6×10^{-8}	3.8×10^{-9}
8.25	0.37	5.6×10^{-9}	3.3×10^{-9}
8.54	0.52	2.9×10^{-9}	3.1×10^{-9}

---

*There is an important restriction here, which is met by all indicators you will study in this lab. You must make sure, through good dilution technique, that all solutions have the same total concentration of indicator.

From this we can calculate an average value of 3.2×10^{-9}. Note that this result may seem imprecise, but it is typical of results for determining an equilibrium constant under general laboratory conditions.

Note that Figure L-2 shows light absorbed at two wavelengths. We can use the same procedure to determine K_a using data from the wavelength of 440 nm.

CAUTION: In this laboratory, you handle a variety of different colored substances. The pleasant colors that result make it easy to forget that they are in solutions with significant, and dangerous, acid and base properties. Be certain to wear proper eye protection at all times.

Procedure

Your eye is a sensitive color detector, although it cannot provide very good quantitative data. You will begin this experiment by using tissue culture plates to determine the values of pH where your assigned indicator undergoes a color change. These will allow you to plan for similar experiments using the spectrophotometer.

You will prepare a series of six solutions for analysis. These will include pH's well above and well below the pK_a for the indicator. You will also measure the spectrum for four solutions at pH's where α varies most significantly—within one pH unit of the pK_a.

Part I: Formation of Groups

Groups of three students are required for this lab. Each student will have a different indicator—bromocresol purple, bromothymol blue, or thymolphthalein. The structures of these indicators in their *acid form* are given in Figure L-4.

Figure L-4

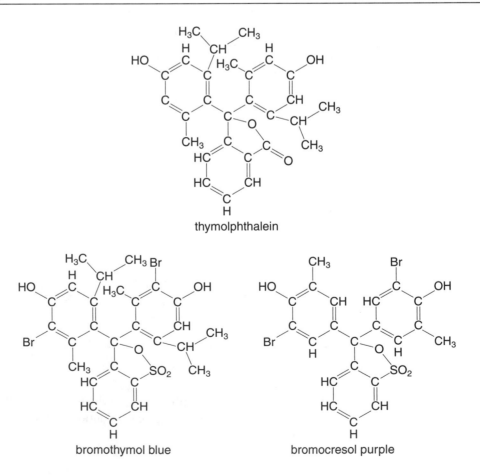

thymolphthalein

bromothymol blue

bromocresol purple

The group should assemble the materials needed for the experiment. One student should be responsible for the tissue culture plates and spectrophotometer cells. The second student should obtain 300 mL of the standard buffer in a 500-mL beaker and 100 mL of NaOH solution in a different clean beaker for the group. The third student should prepare the pH meter and make sure that the group has nine 100- to 250-mL beakers on hand for pH analysis.

Part II: Survey of Indicator Colors over a pH Range (Individual Work)

Each indicator should be tested in the tissue culture plates to determine where it undergoes a color change. This is done by taking the standard buffer and using it to prepare solutions at pH values near ($\pm$0.1 pH unit) 4, 5, 6, 7, 8, 9, 10, and 11. Each student should have a tissue culture plate ready with one or two drops of indicator in each of eight wells.

As a group, take 100 mL of the standard buffer in a 250-mL or larger beaker. Check the pH value. Add NaOH in small amounts (about 0.50 mL). Mix well and record the pH value. When the pH is between 3.9 and 4.1, each student in the group should take some of this for the pH = 4 well in his or her plate.

Continue adding base to the buffer. When the pH is within 0.10 pH unit of each of the target values, each student should again withdraw some buffer for a tissue culture plate well.

The eight solutions in the wells of the tissue culture plate should show where the color changes dramatically over a range of two pH units. This will be your target pH range for the spectrophotometric experiments.

Part III: The Spectrophotometric Titration (Individual Work)

Prepare six clean, dry polystyrene cuvettes. *Use only water to clean these cuvettes. Acetone will dissolve the polystyrene!* Arrange them in a neat row. The cuvettes have two sides that are slightly cloudy and two sides that are clear. Label the cuvettes No. 1 through No. 6 on the cloudy side. Do the same with six small beakers or Erlenmeyer flasks.

Each student will now have to prepare six buffer solutions. One should be at least one pH unit *below* the transition range of the indicator (for example, at pH = 7 for thymol blue). The sixth should be one pH unit *above* the transition range of the indicator (for example, at pH = 10 for thymol blue). The pH of solutions 2 to 5 will span the center of the transition range (for example, because thymol blue changes pH between pH = 7 and pH = 9, a suitable set of values would be pH = 7.5, 7.9, 8.3, and 8.7).

About 150 mL of the buffer should be taken by each student in a 400-mL beaker. Record the pH, then use the NaOH solution to adjust the pH to the lowest pH in the series. Record the pH, then use a volumetric pipet to transfer exactly 25 mL of this solution into the first small beaker or flask.

Next, adjust the pH of the remaining buffer to the lowest pH value for the transition range of the indicator. You should be within 0.10 pH unit of the target value. Record the pH, then use a volumetric pipet to transfer exactly 25 mL of this solution into the second small beaker or flask.

Continue to add NaOH to the buffer to reach a point within 0.10 pH unit of each of the values in the transition range. At each point, record the exact pH and use a volumetric pipet to obtain 25 mL for the third small beaker or flask. Conclude by adding NaOH to reach a pH at least 1.5 units above the highest value of the transition range.

Each student will now have six beakers or flasks containing buffer at the different pH values. Add *exactly* the same amount of indicator solution to each (0.50 mL is probably enough). Mix well, then put some of each into one of the six cuvettes.

Each solution should be analyzed on the spectrophotometer. If a single-wavelength unit is used, set the instrument to the wavelength listed in Table L-1. If a recording spectrophotometer is used, be certain to obtain the absorbance value at the wavelength in the table.

TABLE L-1

Indicator	Wavelength
Bromocresol purple	595 nm
Bromothymol blue	610 nm
Thymolphthalein	610 nm

Report

You should discuss in your report how all members of the group determined the pH range to use to study his or her indicator. Present a table listing the following data for each of the four intermediate samples you examined: pH of the sample, apparent color (that is, what you see), and absorbance at the reference wavelengths.

Prepare a plot of the absorbance values versus pH for your spectra.

A second table should convert the values for the absorbances in cuvettes 2 through 5 into values for α and $\alpha/(1 - \alpha)$ and convert the pH's into $[H^+]$. The last column of this table should have a calculated K_a.

This experiment provides four independently determined values for K_a—one each from samples 2 through 5. Calculate an average and a standard deviation for K_a.

EXPERIMENT 2
Application Lab: Equilibrium of Protein–Ligand Binding

Pre-Laboratory Assignment   **Due Before Lab Begins**

NAME: _____

Complete these exercises after reading the experiment but before coming to the laboratory to do it.

1. A chemist finds that addition of protein to a solution of a substrate causes the absorbance of a solution to change from 1.23 to 0.67. What is ΔA?

2. What is the ideal range for absorbance values, A, in this experiment?

3. Describe how you will determine the best wavelength to study for this reaction.

4. You determine that a mixture of 2.3×10^{-5} M substrate and 1.9×10^{-5} M protein results in the binding of 23% of the substrate. Determine the concentrations of protein–substrate complex, free protein, and free substrate in this mixture. Then determine the equilibrium constant K for the binding of the substrate by the protein.

5. During this laboratory, you will have to determine the reaction of a substrate with albumin, derived from bovine (cattle) blood. If you were working with human protein, what safety precautions would you need to take?

EXPERIMENT 2
Application Lab: Equilibrium of Protein-Ligand Binding

In the skill-building lab, you worked with a system that was responding to an external influence: a dye changing its color as the pH is altered. In that case, we knew the concentration of one component of the equilibrium, the hydrogen ion. We needed to use our data to determine the concentration of the other two components.

A more complicated situation occurs when we *do not know* the concentration of *any* particular component of the equilibrium. This is the case, for example, when zinc ion reacts with lactate ion, $C_3H_5O_3^-$, to form a bound complex. Here *complex* does not mean *complicated*. Rather, it is a chemical term that refers to a *connected system*, where the connection in this case is between the metal and lactate ions. This can be described by an equation for the chemical reaction of zinc and lactate ions:

$$Zn^{2+} \ (aq) + C_3H_5O_3^- \ (aq) \rightleftarrows Zn(C_3H_5O_3)^+ \ (aq)$$

When an equilibrium reaction occurs, macroscopic properties, such as color, stop changing after a period of time. But at the atomic scale, molecules are constantly converting back and forth. Because the atomic-level process occurs equally fast in the forward and backward directions, the net change is zero and there is no observable change at equilibrium.

What happens when we try to keep track of an equilibrium reaction? We cannot assume that the reaction goes to completion, so we cannot calculate how much of any component is present based on simple stoichiometry. Instead, the final concentrations must be experimentally observed quantities. But the balanced equation is still useful to us as a guide to the relative concentration changes.

An experiment in this system might start by preparing a solution that has $[Zn^{2+}] = 0.10$ M and $[C_3H_5O_3^-] = 0.10$ M. These are the initial concentrations. The system is allowed to react, forming some of the product, $Zn(C_3H_5O_3)^+$. When the system reaches an equilibrium state, the concentrations of all the species present are measured. The experiment would then show $[Zn^{2+}]_{eq} = 0.031$ M, $[C_3H_5O_3^-]_{eq} = 0.031$ M, and $[Zn(C_3H_5O_3)^+]_{eq} = 0.069$ M. This is summarized in Figure L-5. Note that we know the starting and final concentrations and that in this case, the change in the concentrations is calculated from the difference between the initial and ending concentrations.

Another example of such a complex occurs when a protein binds a small molecule. Proteins do this with a number of different results, such as the transport of small molecules and ions and carrying signals from cell-surface receptors. Proper binding is essential for normal cell function and cell growth. For example, calcium binding by a protein called protein kinase c is required for the protein's activity, which controls functions such as programmed cell death and cell differentiation.

Complexes between proteins and substrates can be disrupted in a number of ways. One of the most important is the binding of another molecule instead. The other molecule inhibits the binding of the substrate and is called an *inhibitor*. An example of this is the well-known antibiotic penicillin, which binds to a bacterial protein, inhibiting its ability to bind to its normal substrates and function properly.

Figure L-5 Bookkeeping of a reaction that results in an equilibrium mixture. All concentrations are in moles per liter.

	Zn^{2+}	+ $C_3H_5O_3^-$	$\rightleftarrows$	$Zn(C_3H_5O_3)^+$
Balanced reaction:				
Initial concentrations:	0.100	0.100		0.0
Change in concentrations:	−0.069	−0.069		+0.069
Ending or equilibrium concentrations:	0.031	0.031		0.069

Figure L-6 Bookkeeping of a protein/substrate equilibrium

	P	+	S	$\rightleftharpoons$	PS
Balanced reaction:	P	+	S	$\rightleftharpoons$	PS
Initial concentrations:	1.0×10^{-6}		3.6×10^{-6}		0.0
Change in concentrations:	$-x$		$-x$		$+x$
Ending or equilibrium concentrations:	$1.0 \times 10^{-6} - x$		$3.6 \times 10^{-6} - x$		x

The bookkeeping required for a protein–substrate binding is the same as for any other equilibrium reaction. This is shown in Figure L-6. Note that one apparent difference from the zinc–lactate example results from the much lower concentrations that are involved.

$$P \ (aq) + S \ (aq) \rightleftarrows PS \ (aq)$$

Following Equilibrium Reactions

We are left with an important problem in this case: How can we get information on any of the concentrations? There are many ways to approach this problem experimentally. Which one we choose depends on the nature of the components of the reaction. We can use electrochemistry, for example, if one of the components is electrochemically active. Another and more common way is to follow the changes with light, through spectrophotometry.

In the case of a pH-dependent equilibrium, like the one we studied in the skill-building lab, we knew that we had a way of adjusting the equilibrium so that the indicator was essentially all in the acid form or all in the basic form. In the case of protein complexes, we do not have such a simple method. But we can look for some observable signal that the color corresponds to an all bound or an all unbound system.

Let's say that the substrate S absorbs a different amount of light when complexed. We can easily tell how much color is absorbed for the uncomplexed S—we simply take its spectrum when no protein is present. As we add more and more protein, more and more of S is complexed. Eventually, we should be able to add so much protein that essentially all of the S is complexed. We call such a situation saturated, because addition of more protein will not affect the absorbance.

For the experiment today, your group is going to determine the equilibrium of binding of a simple dye, phenol red (all in its red form), with a common protein, albumin. Albumin (also studied in a different way in Experiment Group E) is present in the blood (serum). It has a role in regulating blood pressure and is also a very important protein for transporting smaller molecules around the body.

There are special terms used to refer to the binding equilibria that you will study today. The binding of a substrate to an enzyme is often called an *affinity reaction*. Therefore, the equilibrium constant K for the reaction is called an *affinity constant*.

The algebra that you can use for today's lab is similar to that used for the previous K_a of an indicator lab. You will be studying only one wavelength. There are three cases that are important: when no substrate is bound, when all the substrate is bound, and when some of the substrate is unbound.

When none of the substrate is bound, the absorbance of a solution all comes from the substrate itself. When all of the substrate is bound, we have an absorbance from the bound substrate. And in between both bound and unbound substrate contribute:

None bound: $A = A_{\text{substrate, unbound}}$

All bound: $A = A_{\text{substrate, bound}}$

Some bound: $A = A_{\text{substrate, unbound}} + A_{\text{substrate, bound}}$

Figure L-7 Bookkeeping of a protein/substrate equilibrium

Balanced reaction:	P	+	S	$\rightleftharpoons$	PS
Initial concentrations:	P_{total}		S_{total}		0.0
Change in concentrations:	$-\alpha\, S_{total}$		$-\alpha\, S_{total}$		$+\alpha\, S_{total}$
Equilibrium concentrations:	$P_{total} - \alpha\, S_{total}$		$S_{total} - \alpha\, S_{total}$		$\alpha\, S_{total}$
	$P_{total} - \alpha\, S_{total}$		$(1-\alpha)\, S_{total}$		$\alpha\, S_{total}$

We now consider how the equilibrium reaction will affect the concentrations of P, S, and PS. If a substrate is partially bound, then we can say that a fraction α is bound. In this case, our bookkeeping can look like that in Figure L-7. There is a change in the concentration of unbound substrate, by an amount $-\alpha\, S_{total}$. Because the reaction consumes one molecule of P for every molecule of S, the change in the concentration of P will be the same amount, $-\alpha\, S_{total}$. And the concentration of PS will increase by $+\alpha\, S_{total}$.

Note that the total amount of protein is always equal to the concentration we put into the solution. The total amount of substrate is also a constant. The one variable is the extent of the reaction, α.

In Figure L-7, we have taken our multiple-variable system and reduced it to just one unknown, α. We know P_{total} and S_{total}. To determine α, we compare the light absorbed by the unbound and bound substrate. The absorbance when none of it is bound is $A_{unbound}$. The absorbance when all of it is bound is A_{bound}. The difference between these two is $\Delta A = A_{unbound} - A_{bound}$. If we keep the total concentration of substrate the same in all solutions, then we can determine the fraction bound, α, by comparing ΔA_{max} (when it is all bound) with ΔA at intermediate values:

$$\alpha = \frac{\Delta A}{\Delta A_{max}}$$

Once we know α and the total concentrations of added protein (P_{total}) and substrate (S_{total}), we can calculate the equilibrium concentrations:

$$\text{Concentration of P} = P_{total} - \alpha\, S_{total}$$
$$\text{Concentration of S} = (1-\alpha)\, S_{total}$$
$$\text{Concentration of PS} = \alpha\, S_{total}$$

Putting these concentrations into the equilibrium expression yields a value for the equilibrium constant, K.

$$K = \frac{PS}{P \times S} = \frac{\alpha\, S_{total}}{(P_{total} - \alpha\, S_{total})(1-\alpha)S_{total}} = \frac{\alpha}{(P_{total} - \alpha\, S_{total})(1-\alpha)}$$

Procedure

Part I: Formation of Groups

Groups of two will be responsible for testing the albumin–phenol red reaction and each group member will prepare three solutions for the primary equilibrium curve study.

Part II: Determining the Concentration Range for Equilibrium Study (Group Work)

You need to know the right set of concentrations of protein for the study. You should use 10 wells of a tissue culture plate to examine the different concentrations. The easiest way to do this is to start with 15 drops of stock protein in the lower left well.

Then take 5 drops of this and place it in the next well. Add 10 drops of buffer. This dilutes the original mixture by 33% to 67% of the starting value. Do this again and again across all 10 wells. If you do this across all 10 wells, you will have achieved a total dilution of 0.67^9. The final well will have less than 3% of the starting concentration. You should then add the same amount of substrate to each well. Three drops should work well.

Examine the different wells. Each has a constant concentration of substrate and a different concentration of protein. Where are the color changes the most significant? This is the major range for the study of the different protein concentrations.

Part III: Spectrophotometric Data Collection (Group Work)

In the skill-building lab, you were able to follow the change of the absorbance over a pH range. Here you have to study the change over a range of different initial protein concentrations. To do this, use your data from the tissue culture plates. Use these concentrations to prepare a series of 10.0-mL samples with the same substrate concentrations and the proper range of protein to show the change. You should measure at least six different concentrations, including a sample with no protein.

The last step is to determine the concentration when the system is saturated. You should prepare three other solutions, each with the concentration of substrate that you chose. They should also have 10, 50, and 100 times as much protein as any of the original set of six cells. Measure the absorbance of each of these.

Part IV: Disruption of Protein–Ligand Binding (Group Work)

Other small molecules can compete with a substrate for binding to a protein. These inhibitors can be critical in disrupting the transport of molecules in an organism and within a cell. You will not have to determine the effect of a small molecule quantitatively. But you will be given a solution of a molecule (sodium salicylate) that you will add to each of your spectrophotometric samples. Add the same number of drops, mix well, and then determine the absorbance of each sample.

Report

You have just carried out data collection to determine the fundamental behavior of a protein and its substrate. You should prepare a report that discusses the equilibrium from several different perspectives, using the following format, as appropriate for the scenario in which you are investigating the parameters of a new drug transport mechanism.

- Description of the materials used.
- Analysis of the binding system to ensure accurate determination of A_{bound} and $A_{unbound}$.
- Tabulation of substrate and protein calculations along with observed A values and calculated values of α.
- Determination of K at different concentrations of protein and determination of an average value.
- Discussion of the disruptive effects of sodium salicylate.
- Discussion of whether albumin in the blood will transport the substrate well. You may assume a blood albumin concentration of 1.0×10^{-5}M.

Experiment Group **M**

Gases: Collect Hydrogen Gas for Potential Use in a Fuel Cell

Photovoltaic cells like this one are important in the solar/hydrogen economy (PhotoDisc.)

Purpose In this experiment group, you will begin by studying the properties of gases and then you will focus on a specific gas—hydrogen. In the skill-building lab, you will examine the relationships among pressure, volume, temperature, and number of particles for different gases. You will also learn how to collect a gas generated in a chemical reaction. In the application lab, you will study conditions that optimize the production of hydrogen gas from the electrolysis of water using a solar panel. You will also investigate the use of hydrogen as a fuel in fuel cells.

Schedule of the Labs

Experiment 1: Skill-Building Lab: The Gas Laws—A Study of the Behaviors of Gases

- Determine the relationship between pressure and volume at a constant temperature for different gases (group work).
- Determine the relationship between the volume and temperature of a gas at a constant pressure (group work).
- Determine the relationship between the pressure and temperature of a gas at constant volume (group work).
- Determine the value of the universal gas constant (group work).

Experiment 2: Application Lab: Hydrogen—A Clean Fuel

- Design an apparatus for the collection of hydrogen gas generated by electrolyzing water (group work).
- Determine the relationship between the power output of a solar panel and the angle of incident radiation (group work).
- Determine what other conditions optimize the production of hydrogen gas from electrolysis of water (group work).

Scenario The automobile is generally considered a necessity in the United States and other developed countries. It is the primary mode of transportation for millions of people. Most cars still run on gasoline, a nonrenewable resource that, when burned in an internal combustion engine, emits carbon dioxide, carbon monoxide, unburned hydrocarbons, nitrogen oxides, and sulfur oxides. While catalytic converters help to clean up these emissions, automobile emissions are still polluting the atmosphere and contribute to environmental problems such as global climate change, ground level ozone, and acid rain.

Engineers have designed cars that run on fuels other than gasoline. One such fuel is hydrogen gas, which can be burned in an internal combustion engine, producing only water vapor and some nitrogen oxides. These latter come from the N_2 and O_2 in the atmosphere reacting at the high temperatures in an internal combustion engine rather than from the fuel. However, a cleaner way to utilize hydrogen as a fuel is in fuel cells. Here, hydrogen and oxygen are combined electrochemically, without a flame, to produce electricity, heat, and pure water—no polluting emissions. So why aren't fuel cells used in most automobiles now? Three very important questions need to be addressed before hydrogen and fuel cells become commonplace.

First, what will supply the energy to make hydrogen gas from water in the first place? The hydrogen can be made by refining fossil fuels, but in that case the fuel is still nonrenewable and CO_2 emissions will still occur. If this energy comes from the sun, this represents a clean, renewable system, but the question then becomes "Is solar energy a viable alternative?" Second, how will hydrogen gas be stored and transported economically and safely? Finally, can fuel cells be economically produced as cheap power cells for cars and other uses?

In this experiment group, you will be a part of the research team studying hydrogen as a fuel. You will use a photovoltaic (also known as solar) panel to decompose water into hydrogen and oxygen gas. This gas can then be converted back into water, while also producing electricity, in a fuel cell. You will use the results of your research to begin to answer the questions posed above.

EXPERIMENT 1
Skill-Building Lab: The Gas Laws—A Study of the Behaviors of Gases

Pre-Laboratory Assignment   **Due Before Lab Begins**

NAME: _____

Complete these exercises after reading the experiment but before coming to the laboratory to do it.

1. Differentiate among solids, liquids, and gases on a molecular level.

2. Calculate the pressure of a sample of gas collected over water at 25 °C if the atmospheric pressure is 0.948 atm. 1 atm = 760 mm Hg.

3. The vapor pressure of water at 16 °C is 13.6 torr and it is 15.5 torr at 18 °C. What is the vapor pressure at 17 °C?

4. Write the balanced equations for the reaction between magnesium metal and hydrochloric acid solution and for the decomposition of hydrogen peroxide into water and oxygen gas.

5. A graph of a set of data reveals a linear relationship between the two variables. The best-fit equation is $y = 0.0254x + 64.938$. Determine the value of the x-intercept for this line.

6. If the water levels on both sides of the apparatus shown in Figure M-1 are equal, how does the pressure exerted by the gas on the left compare with the pressure exerted by the atmosphere on the right?

7. What special precautions are required in setting up a dry ice/alcohol bath? Explain.

Figure M-1 Setup for collecting gas in a eudiometer

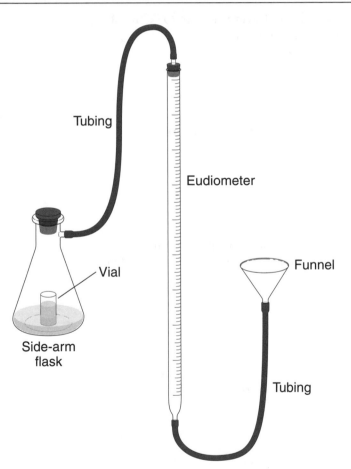

EXPERIMENT 1
Skill-Building Lab: The Gas Laws—A Study of the Behaviors of Gases

Background

The gaseous state is one of the three common states of matter. Gases possess properties that distinguish them from liquids and solids on a macroscopic level: Gases exert pressure in all directions, they are compressible, they fill the volume and take on the shape of their containers, they have relatively low viscosity and density (under normal conditions), and they are miscible. Gases can also be distinguished from solids and liquids on a molecular level: According to kinetic molecular theory, ideal gases are described as individual particles in constant random straight-line motion. When they collide with each other or with the walls of their containers, the collisions are completely elastic; that is, there is no net loss of kinetic energy. The particles do not exert any intermolecular forces of attraction, and the volume of the particles themselves is negligible compared with the volume of their container. Real gases behave very much like ideal gases at high temperatures and low pressures.

Several empirical laws describe the behavior of gases. You will be investigating some of these gas laws in this experiment. One of the laws you will use, but not specifically explore, is Dalton's law of partial pressure. This law states that the total pressure of a mixture of gases is equal to the sum of the partial pressures of each of the gases:

$$P_T = P_A + P_B + P_C + \cdots$$

where P_T is the total pressure exerted by all gases in the mixture and P_A, P_B, P_C, etc. are the pressures that each of the component gases exert. When a gas is collected over water, water vapor is also present in the container. Therefore, the total pressure of the gaseous sample is the sum of the pressure of the gas under study and the vapor pressure of water, $P_T = P_{\text{dry gas}} + P_{\text{water vapor}}$. The vapor pressure of water varies with temperature. Tables of the vapor pressure of water at different temperatures are available in appropriate handbooks. Then, if the total pressure and the vapor pressure of water are known, you can calculate the pressure of the gas being studied.

CAUTION: Several parts of this experiment involve high temperatures or very low temperatures. All apparatus appear the same as at room temperature. Be very careful not to burn yourself.

Procedure

Part I: Formation of Groups

You will work in groups of two or three people. You must all participate in the experiment, record data in your own notebooks, and write up your own reports. You will need to work together in each part of the experiment in order to collect data efficiently.

Your instructor may assign only portions of this experiment for your group to complete.

Part II: Relationship Between Volume and Temperature: Charles' Law* (Group Work)

Set up this experiment right away and monitor it throughout the lab period as you continue with other experiments. *Do not leave the hot plate unattended.* Once the

---

*The procedure for the Charles' law experiment follows the procedure outlined in Kim, M-H, Kim, M. S., and Ly, S-Y, *J. Chem. Educ.*, **2001**, 78, 238–240.

system is removed from the heat, you can return at regular intervals to record the necessary data.

1. Fill a 1-L tall form beaker with about 900 mL of water.
2. Fill a 10-mL graduated cylinder with about 7 mL of water. Seal the opening of the graduated cylinder with your finger and invert it into the tall form beaker. Record the volume of the trapped air sample in the graduated cylinder when it is submerged in the 1-L beaker of water.
3. Place the beaker on a hot plate, position the thermometer in the water in the beaker, and heat until the temperature reaches about 70 °C.
4. Remove the beaker from the hot plate, reposition the thermometer if necessary, and record the temperature of the water and the volume of the air trapped in the inverted graduated cylinder.
5. Allow the system to cool. Measure the temperature of the water bath and the volume of the trapped air at regular intervals. Slowly begin to add ice to the water bath once the temperature cools to about 40 °C and continue recording temperatures and corresponding volumes of the trapped air.
6. Record the value for the barometric pressure.

Part III: Relationship Between Pressure and Volume: Boyle's Law (Group Work)

You will use a digital system to collect data. The pressure sensor will record the pressure of the gas in the syringe and you will read the volume directly from the syringe.

1. To prepare for collecting pressure–volume data:
   - Plug the pressure sensor into the digital system as specified by your instructor.
   - Connect the syringe to the three-way valve on the pressure sensor by gently twisting the syringe until it locks in place. If your pressure sensor has a valve, attach the syringe to the valve in line with the tube coming out of the pressure sensor.
   - Open the side-arm of the pressure sensor valve to allow air to enter and exit. If your pressure sensor has a valve, align the valve so it detects pressure in the syringe.
   - Move the piston so the bottom of the black ring on the plunger is aligned with some marked volume.
2. Turn on the digital system and calibrate for atmospheres.
3. To collect data:
   - Record the pressure at the initial volume.
   - Move the piston to position the front edge of the inside black ring at some new volume on the syringe. Hold the piston firmly in this position until the pressure value stabilizes.
   - When the pressure reading has stabilized, record the pressure and the volume.
   - Repeat the procedure for at least another six volume measurements between 6.00 and 20.00 mL.
   - When you have finished collecting data, examine the data points along the displayed graph of pressure versus volume. Record the pressure (round to the nearest 0.01 atm) and volume data pairs in your lab notebook. Export the data to print the graph if possible.

4. Fill the syringe with a different gas and repeat the experiment. Consult with your instructor to find out what other gases are available.

Part IV: Relationship Between Pressure and Temperature (Group Work)

Obtain a rigid, sealed container equipped with a pressure gauge. Record the room temperature and the initial pressure of the gas trapped in the container. Place the apparatus in a dry ice/alcohol bath, then in an ice bath, and finally in a boiling water bath. Be sure to leave the apparatus in each bath for 10–15 min so the gas can reach the temperature of the bath. Record both the pressure and temperature in each case.

Part V: Relationship Between Volume and Moles: Avogadro's Law* (Group Work)

1. Assemble the apparatus as shown in Figure M-1. Connect the funnel to the end of the eudiometer with a 75-cm-long piece of latex tubing. Place a one-hole stopper at the top of the eudiometer and position a Pasteur pipet through the hole in the stopper so that the tip of the pipet is below the top of the eudiometer. Attach one end of a 50-cm piece of tubing to the wide end of the Pasteur pipet and attach the other end to the arm of the side-arm flask. Remove the rubber stopper from the top of the eudiometer and fill the eudiometer and tubing with water. Add enough water so that the eudiometer and the stem of the funnel are almost filled. Be sure there are no bubbles in the tubing.

2. Measure the air temperature in the room and the barometric pressure.

3. Pipet 3.0 mL of 3.0 M HCl into a clean vial.

4. Polish a strip of Mg metal. Cut an approximately 4-cm piece from the strip and weigh it (± 0.001 g). Place the piece of Mg in the flask.

5. Carefully lower the vial of HCl into the flask and stopper the flask.

6. Place the stopper on the eudiometer, adjust the height of the funnel so that the water level in the funnel is even with the water level in the eudiometer, and record the initial volume reading on the eudiometer.

7. One member of the group must hold the stopper on the eudiometer while another carefully shakes the flask to tip the vial. A reaction will start when the acid and magnesium are in contact.

8. When there is no more evidence of a reaction, let the apparatus sit for 15 min to allow the gas to cool to room temperature.

9. Reposition the funnel so that the water level in the funnel is again level with the water level in the eudiometer. Record the final volume reading on the eudiometer.

10. Dispose of the contents of the flask in an appropriate waste container and rinse the flask. Run at least three trials.

11. Repeat the experiment, but this time collect oxygen gas from the decomposition of hydrogen peroxide. Record the concentration of the hydrogen peroxide solution. Pipet 2.5 mL into the flask and fill the vial with catalase solution. Catalase is a catalyst that will increase the rate of decomposition of the hydrogen peroxide. Carefully lower the vial into the flask and seal the flask. Repeat steps 6–10. Run at least three trials of this reaction.

---

*The apparatus for the collection of a gas in a chemical reaction is based upon a design presented in Moss, D. B., and Cornely, K., *J. Chem. Educ.*, **2001**, *78*, 1260–1261.

Analysis

Part II

Consult an appropriate reference, such as the *CRC Handbook of Chemistry and Physics*, and record the vapor pressure of water at each temperature. If the vapor pressure at a particular temperature is not listed on the table, interpolate between the values listed for the temperatures just above and just below the experimentally measured temperature. Assume a linear relationship exists between vapor pressure and temperature. This is a valid assumption over very narrow temperature ranges.

Because you are studying the relationship between the volume and temperature of the trapped gas, you need to correct each measured volume for the presence of the water vapor. While the number of particles of the original trapped air sample remains constant, the number of water molecules present changes with temperature. At higher temperatures, more water molecules are present in the vapor phase. The pressure exerted on the trapped gases remains constant. (We are assuming this is atmospheric pressure. While the gases are also under some additional pressure due to being below the liquid level in the outer beaker, that pressure is assumed to be negligible compared with atmospheric pressure in this experiment and it will be ignored.) Therefore, the volume must increase to accommodate the additional water molecules. The partial volume of the dry air is given by

$$V_{air} = V_{total}\, X_{air}$$

where X_{air} is the fraction of the air molecules in the total volume. Because the pressure exerted by a gas is directly proportional to the number of gaseous particles, the fractional abundance of air (X_{air}) can be determined by a pressure ratio

$$X_{air} = P_{air}/P_{total}$$

where P_{total} is equal to barometric pressure and $P_{total} = P_{air} + P_{water}$. Then

$$V_{air} = V_{total}[1 - (P_{water}/P_{total})]$$

Prepare a graph with the volume of the dry air along the y-axis and the temperature along the x-axis. Plot the volume and temperature data. Use a spreadsheet or graphing calculator to determine the best-fit line. Find the x-intercept. Post your value for the class.

Part III

Prepare a graph with pressure along the y-axis and volume along the x-axis. Plot the pressure–volume data for each gas used.

Part IV

Plot the pressure–temperature data using a graphing calculator or spreadsheet. Obtain the best-fit equation. Determine the value of the x-intercept.

Part V

For each trial, calculate the moles of gas produced using the moles of limiting reactant and the balanced chemical equation for the reaction. The HCl solution (and the H_2O_2) displaced some air in the system. Then the gas produced in the reaction displaced some water in the eudiometer. Therefore, to determine the volume of the gas generated in the reaction, subtract the initial volume in the eudiometer and the volume of the solution from the final volume in the eudiometer. The purpose of the funnel is to equalize the pressure of the gas to that of the atmosphere by moving the funnel until the water level in the eudiometer is equal to the water level in the funnel. However, because the gas was collected over water, the total pressure is due to both the gas generated in the reaction and water vapor. Look up the vapor pressure of water at room temperature and calculate the pressure of the "dry" gas using Dalton's law of partial pressures. Tabulate the pressure, volume, temperature, and number of moles of gas for all trials.

Calculate the molar volume of the gas produced in each trial. Tabulate these values.

Report

You explored the relationship between pressure, volume, temperature, and number of particles (moles) of gases in this experiment. Post your data from each section for the entire class and discuss the class results. Summarize the results from each part of this experiment, noting the variables that were manipulated and those held constant. Then answer the following questions in your report.

1. How did the value of the x-intercept in the volume–temperature experiment compare with the x-intercept in the pressure–temperature experiment? What is the value for y at the x-intercept? What physical property does y represent in each of these two experiments? Reset the x-axis so that the x-intercept is now zero, but keep the magnitude of each degree the same. What physical property does x represent in each experiment? State the values of the properties represented by y in each experiment when x is zero on the new axis.

2. Was the relationship between pressure and volume dependent upon the identity of the gas used? Explain.

3. Use your results from Part V to determine an average value for R in the ideal gas law, $PV = nRT$. Do your results and the class results support the idea that R is a universal gas constant? Explain.

4. According to the ideal gas law, how are P and V related at constant n and T? How are V and T related at constant n and P? How are P and T related at constant V and n? How are V and n related at constant P and T? Do your data support the relationships illustrated in this equation? Explain. Include examples from your results in your explanation.

EXPERIMENT 2
Application Lab: Hydrogen—A Clean Fuel

Pre-Laboratory Assignment **Due Before Lab Begins**

NAME: _____

Complete these exercises after reading the experiment but before coming to the laboratory to do it.

1. Calculate the number of moles of H_2 gas in an 80.5-mL sample collected at 25 °C and 742 mm Hg.

2. Write the balanced chemical equation for the decomposition of water.

3. We find that the decomposition of water produces 35.4 mL of dry H_2 at 28 °C and 0.985 atm. How many moles of H_2O decomposed? How many milliliters of dry oxygen will be formed, assuming the same temperature and pressure?

4. What special precautions must you take while handling hydrogen gas? Explain.

EXPERIMENT 2
Application Lab: Hydrogen—A Clean Fuel

Background

This lab will give you experience in different ways of capturing, storing, and using energy. High-quality energy is found in sunlight, for example. It is readily available, it is free, and it represents a source of energy external to the earth. However, to use the energy in sunlight, we must convert the sun's light energy into another form of energy. In this lab, the conversion will be done through the use of an electrical current that allows us to generate hydrogen and oxygen gas from water. The hydrogen and oxygen gases possess an easily transported form of energy: chemical energy. Then we cycle the hydrogen and oxygen back into water and release the energy as electrical energy in a fuel cell. Eventually, this energy becomes the lowest quality of energy, heat. Thus, the center of this lab is a chemical cycle among water, hydrogen, and oxygen, all driven by the energy of the sun.

 CAUTION: Hydrogen gas is flammable. Keep the reaction system involved in the production of hydrogen gas away from any ignition source. Use a very small sample of gas, less than 5 mL, when testing its identity with a burning splint.

Procedure

Part I: Formation of Groups

You will work in groups of two or three people. You must all participate in the experiment, record data in your own notebooks, and write up your own reports. You will need to work together in each part of the experiment in order to collect data efficiently.

Part II: Electrolysis of Water (Group Work)

You will work with your group to design a system to collect and measure the volume of the hydrogen gas produced by the electrolysis of water. You will use a photovoltaic (PV) panel to provide the electricity to decompose the water. Obtain two platinum electrodes and attach one to each lead on the PV panel. Fill a 250-mL beaker about three-quarters full of deionized water. Add enough Na_2SO_4 to bring the concentration to about 1 M. Insert the platinum electrodes in the water, expose the PV panel to light, and collect a small sample of gas at each electrode. Identify which gas is hydrogen and which is oxygen. Label the electrode at which the hydrogen is produced, since for the remainder of the experiment you are only interested in monitoring the production of hydrogen gas.

Set up an apparatus that will enable you to collect hydrogen gas. Add a multimeter to the system so that you can also record voltage and current. Your task is to change variables in order to determine what conditions are optimal for maximizing the production of hydrogen gas. You will be altering light sources, angles of incident radiation, electrode material, and any other variables agreed upon by your team and approved by your instructor. Be sure to control all variables but the one under study. You will need to measure the volume of hydrogen generated in a specific period of time. Keep the time constant from trial to trial in order to make comparisons. In addition to overall volume collected and total time, measure the volume, time, temperature, voltage, and current every 10–30 s. Record the room temperature and the barometric pressure as well.

Once you have determined the optimum conditions for generating hydrogen, use a fuel cell to run a fan or light bulb.

Report

Plot the volume of hydrogen gas produced as a function of the angle of incident radiation. Describe the relationship illustrated by the graph. Summarize all of your results and recommend the optimal conditions for the production of hydrogen gas from the electrolysis of water.

Plot PV versus nT for each trial in which the angle of incident radiation was changed. What is the slope of the line? How does this compare with the value for R in the ideal gas law? Account for differences.

Also include answers to the following questions in your report.

1. What is the heat of reaction for $2 H_2 (g) + O_2 (g) \rightarrow 2 H_2O (g)$?
2. What would be the volume of hydrogen gas at STP needed to provide the same amount of energy as 20 gal (64 kg) of octane (gasoline)?
3. How might one reduce this equivalent volume?
4. Have you accounted for the small amount of water vapor mixed with the H_2 gas due to water's partial pressure? There is a small correction needed here as well.

Finally, use your experimental results, along with information from at least three other sources, to tackle the questions (outlined in the scenario) that need addressing before hydrogen fuel cells can become commonplace.

Index